Building Information Modeling

To Jean-Louis Le Moigne, this book on modeling and systems engineering. A story of handover and engineers, of academics and business practitioners working together.

SCIENCES

Mechanics, Field Director – Gilles Pijaudier-Cabot

Civil Engineering and Construction,
Subject Head – Fabrice Gatuingt

Building Information Modeling

Shared Modeling, Mutual Data, the New Art of Building

Coordinated by
Régine Teulier
Marie Bagieu

iSTE

WILEY

First published 2023 in Great Britain and the United States by ISTE Ltd and John Wiley & Sons, Inc.

ISTE Ltd
27-37 St George's Road
London SW19 4EU
UK

www.iste.co.uk

John Wiley & Sons, Inc.
111 River Street
Hoboken, NJ 07030
USA

www.wiley.com

Library of Congress Control Number: 2023932247

British Library Cataloguing-in-Publication Data
A CIP record for this book is available from the British Library
ISBN 978-1-78945-110-8

ERC code:
PE8 Products and Processes Engineering
PE8_3 Civil engineering, architecture, maritime/hydraulic engineering, geotechnics, waste treatment
PE6 Computer Science and Informatics
PE6_3 Software engineering, operating systems, computer languages

Contents

Foreword

Arto KIVINIEMI
School of Architecture, University of Liverpool, United Kingdom

Although there are plenty of books about building information modeling (BIM), it is always a pleasure to see a new book especially when it is written by such a competent group of authors as this book. As with any rapidly evolving technology, BIM is a complex issue and there is not even one commonly accepted definition what BIM is or how we should teach it in academia or use it in industry. Thus, it is important to present different views on the topic. My personal view is that there are still a lot of misunderstandings both in the AECOO (architecture, engineering, construction, owning and operating) industry and academia about the nature and impact of BIM. It is often seen just as a software or technical issue, and in academia it is still too often missing from the curricula or it is more software training than an integral part of the education. In addition, the emphasis is too often on modeling, although it should be on the information content and managing and sharing information. Computer-aided design changed the processes very little as it just automated drafting, but BIM is fundamentally changing the way we design, construct and manage our buildings and built environment and collaborate in projects. BIM is affecting the tasks and processes of the many professions involved in the different phases of the lifecycle of buildings and other parts of our built environment. Although today BIM is mainly used for design and construction, it is indeed an information management tool and sharing platform for the whole lifecycle from the setting of client requirements to the demolition of the building.

I highly recommend this book for all the AECOO professionals as well as academics educating the new professionals and students who need to understand the future of our industry.

Preface

Christophe CASTAING[1,2,3]
[1] *buildingSMART International, Paris, France*
[2] *Groupe Egis, Paris, France*
[3] *Projet MINnD, Paris, France*

In memory of Richard Petrie
Chief Executive Officer of buildingSmart International

P.1. The history behind this volume

This BIM volume of the encyclopedia "SCIENCES" has a history. It was largely written by editors working in academics and the field of construction, under the framework of the national project[1] MINnD (Modeling Interoperable Information for Sustainable Infrastructure[2]). The approach transcribed here is based on an expression of needs, linked to the practice of concurrent engineering in infrastructure contracts for large road, rail or industrial infrastructures, developed initially in the COMMUNIC research project (2007–2011) funded by ANR.

1 A national research project specific to the French Ministry of the Environment, Energy and the Sea, labeled and agreed upon by the latter, and which implies that several companies in the sector are strongly committed.

2 The term "infrastructure", in this context, is used for all of the construction on the territory, including utilities, facilities and manufactured systems such as tolling. The term "public works" could be used.

Building Information Modeling,
coordinated by Régine TEULIER and Marie BAGIEU. © ISTE Ltd 2023.

We can summarize the needs resulting from this practice of concurrent engineering as constituted around the search for gains:

– productivity in data exchanges between the various parties involved, beyond the constraints imposed by the traditional organization of contracts;

– productivity in the validation processes of deliverables at each stage of the lifecycle;

– in the verification of the performance achieved, given the growing importance of requirements, particularly in the environmental field.

It should be noted that the identification of these needs was accompanied by the observation that the software industry evolution did not offer the corresponding technical answers. In particular, it did not offer an answer to this double objective: representation of the works to be built and representation of the processes to ensure the data continuity.

P.2. A heuristic

The process for introducing BIM in the public works sector of construction has generated the following heuristic: a long study within the framework of the COMMUNIC research project on the modeling needs for concurrent engineering processes, which led to a focus on the process's modeling with the initial PAS 1192 British standards and then, naturally, to participate in the work of ISO 19650. The process's modeling was accompanied by, alongside the creation of the PN MINnD, a choice for the appropriation of the infrastructure modeling and the participation in the development and the extension of the conceptual models resulting from the IFC or from GML. For the needs of concurrent engineering, modeling an infrastructure project, rather than a single isolated structure, required a founding statement. It was necessary to introduce the knowledge of systems engineering to understand how to correctly decompose works and processes with two key concepts: that of a system of systems applied to an infrastructure project and that introducing the dichotomy between the system "to be made" and the system "for making". However, it appeared that systems engineering did not solve two problems specific to the description of infrastructure and territory:

– the need to introduce a new and additional decomposition with the spatial de-composition understood as a system or not;

– the interweaving of the decomposition into systems of objects that do not enter into the systemic representation (geology for example), but concomitant to the concept of system of systems.

Next, it appeared that the systems engineering made necessary the appropriation of models engineering to build a stable environment for the representation of conceptual models beyond the BIM (Geographical Information System, task decomposition, WBS) posing the need for an appropriation of STEP 239. Finally, by advancing in the phases of appropriation and the implementation of data modeling, in order to complete the overall approach, it was necessary to enter the pathway of knowledge modeling, around the ISO 12006 and 23386 standards to return the conceptual models to the trades by verifying their ability to carry the semantics related to the uses. It would be interesting to study how much this approach (ISO 8000) is inherited by the knowledge domain developed by the structuralism of language learning.

P.3. Knowledge pathways...

The collaborative exploration of the research theme developed in COMMUNIC, and then in MINnD followed two pathways that led to the accumulation of knowledge.

The first pathway was coming through the appropriation of prior knowledge about ISO 8000 and the standardization work within the framework of buildingSMART International.

As shown in Figure P. 1, taken from the ISO website, ISO 8000 identifies three key stages in the modeling of information exchange: the semantics' description with the data dictionaries development, the description of information exchange needs and the implementation for dedicated uses from a global schema of the infrastructure to be built or maintained. The description of this conceptual scheme is emerging from the knowledge modeling including the semantics defined in the data dictionaries and to be attached to the objects or components under design or under construction.

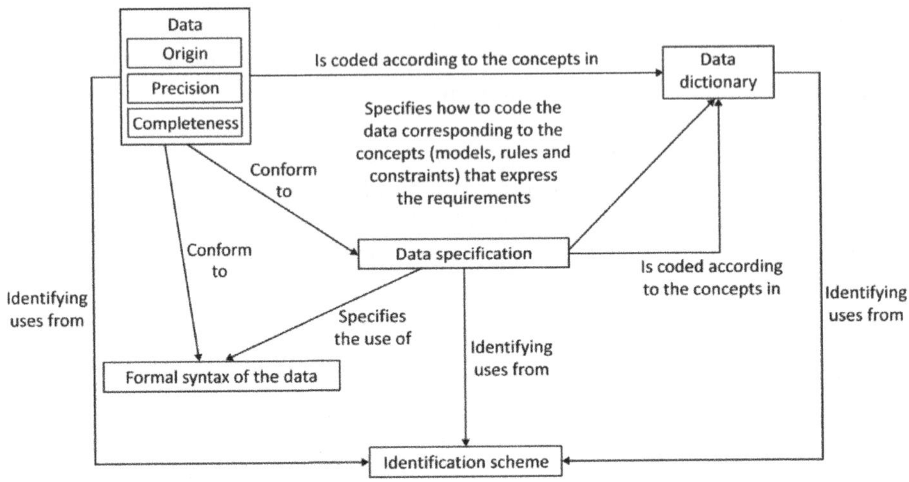

Figure P.1. *The three stages of information exchange (ISO 8000)*

The second pathway is through the appropriation of academic knowledge made available through model engineering and systems engineering developed for the needs of other industries. It was necessary to learn from the concepts of systems engineering in order to escape from the intuitive representation of public works by a tree and to reach the double understanding of a system of systems versus a complex system. It was also necessary to explain and distinguish between the modeling of the system to be delivered (the asset information model) versus the modeling of the system necessary for creating the system to be delivered (the project information model). The results are the pathways of digital continuity to describe a heterogeneous and non-continuous knowledge (a system of systems), which has few equivalents in other industries, but which allows a dialogue on and even offers a solution to other industries.

We must also add the interweaving of new modeling methodologies and technologies coming from the world of geographic information, with the Semantic Web[3].

The construction sector was seen as an archaic sector in its digital transformation. Due to these works and to the mobilization of experts in these fields, it is, now, appearing as the core of a digital twin approach,

3 Available at: www.w3.org.

allowing representation not only of the built (constructed) universe, but also its environment made of objects (non-human being) which interfere with human activity, even if they do not result from it.

P.4. The current challenges...

The maturity acquired by the construction industry in the appropriation of the scientific achievements of digital technology is just blossoming when the need for proactive modification of the built environment and its protection (low carbon strategy) is emerging as the major issue for this century. We can formulate the technical challenges in the following way:

– Will the digital twin capable of representing these two environments (built environment and natural environment) be able to be part of the solution of the low-carbon trajectory?

– While these challenges will force countries and sectors to invest in huge financial efforts, will the construction sector be able to keep control of the tools for the transformation of this built environment, to offer solutions through these new tools?

This other part of the question is by no means secondary. The digital age, which is expanding with the health crisis, is inexorably leading to a transformation of the processes' organization. New practices are being generated, and new sources of revenue will be coveted beyond the traditional circle of the construction ecosystem. For instance, the connected infrastructures will be at the center of a huge market for the Internet of Things.

The acquisition of the mastery of modeling tools of the built universe, including the built and no built environment, allowed an update of a potential open[4] digital economy, controlled by the traditional actors. This ecosystem, which brings together traditional players and the open digital economy, should make it possible to aggregate the emerging technologies of Industry 4.0.

The challenge is therefore twofold:

– on the one hand, construction keeps its hand on the technical specifications of this ecosystem, participates in its design and control, and

4 "Open" refers to the developments of OpenBIM around buildingSMART International, Open GIS around OGC and open source.

then takes its place in the control of the digital continuity throughout the lifecycle of the works and the data;

– on the other hand, the digital industry enters this open ecosystem by accepting not to become the master of the data and its lifecycle.

P.5. The construction ecosystem as an inter-organizations construction

A project such as MINnD allows the French construction industry, with an estimated revenue of 310 billion euros per year (share of GDP), to express a common need around the digital continuity of its data and processes, but also, because of the pooling of means and resources, to move forward in achieving a certain number of these objectives by:

– joint action in AFNOR (CEN and ISO) standardization committees, via professional unions or federations such as FNTP and Syntec-Ingéniérie;

– joint action in learned societies such as buildingSMART France and buildingSMART International, to advance the most technical parts of these works and establish links with international players.

Construction is a key sector in the low-carbon trajectory, which will also involve technical specifications for built objects, materials, processes and, above all, their operation and use. During international economic forums, in Europe or Asia, strategic objectives and interests are clearly expressed: it is no exaggeration to say that the development of OpenBIM type standards, such as the IFC for infrastructure, is part of the strategic objectives of some countries.

March 2023

Introduction

Régine TEULIER[1,2] and Marie BAGIEU[3]
[1]*CNRS, CRG, Institut polytechnique, Paris, France*
[2]*Projet MINnD, Paris, France*
[3]*ESITC Caen, France*

I.1. Genesis of building information modeling

The evolution of computer-aided design (CAD) and the development of computing capacity in the 1970s led to 3D modeling (Baba and Nobeoka 1998). Parametric modeling tools, based on object-oriented programming, became widespread throughout the industry, but their adaptation to the architecture, engineering and construction (AEC) sector took several decades (Penttilä et al. 2007).

I.1.1. *Concept*

Eastman et al. (1975) introduced the concept of building description systems (BDS) in the 1970s and proposed a prototype for developing a general building description system. BDS has since been recognized as the precursor to building information modeling (BIM) tools, with several intermediate steps leading to the BIM process as we know it today (Latiffi et al. 2014), in the following sequence:

– GLIDE: *Graphical Language for Interactive Design* (Eastman and Henrion 1977);

– BPM: *Building Product Model* (Bjork 1989);

– GBM: *Generic Building Model* (Eastman and Siabiris 1995).

I.1.2. *Acronym*

Aish (1986) was the first to describe a CAD system in which information about the structure to be built would be integrated into the design, in a coherent and coordinated manner, by an entire multidisciplinary design team. In this regard, he spoke of building modeling. Later, Van Nederveen and Tolman (1992) proposed to store useful working information in a model, so they introduced the building information model. They wanted to remedy the construction practice characterized by an unstructured organization of the various actors involved in the building, each playing a specific role and having only a specific view of the project data.

More recently, the National BIM Standard – United States® (NBIMS) has defined the term BIM in three ways: building information modeling, model, or management (BIM3) (NBIMS 2015).

I.1.3. *Definition*

For this work, we have chosen as our sole meaning "building information modeling" and refer to the definition given in ISO 19650 (2018). By this definition, BIM is understood as the use of a shared digital representation of a "built asset"[1] to facilitate the design, construction and operation processes, and form a reliable basis for decision-making.

I.2. Presentation of the chapters of this book

In Chapter 1, Régine Teulier and Marie Bagieu examine the economic stakes of BIM. They propose to characterize the nature of BIM as an innovation for the construction sector and methodically examine what constitutes a technological breakthrough, and the consequences associated with such a breakthrough. They propose to retain the hypothesis of qualifying BIM as a technological breakthrough. Next, they consider the different assessments that have been proposed to evaluate the maturity levels

1 Built assets include, but are not limited to, buildings, bridges, roads, process plants.

of BIM in companies, and also show that this type of maturity assessment is now related to the new standards and to the adoption of OpenBIM.

In Chapter 2, Benoît Eynard, Matthieu Bricogne, Alexandre Durupt and Julien Le Duigou retrace, for mechanical engineering and the manufacturing industry, the evolution of multidisciplinary digital design, which occurred about two decades ahead of the evolution in civil and urban engineering and which sheds light on its fundamentals and allows us to question it through hindsight. The authors clarify each concept, tool or method and give precise and stabilized definitions, by situating them in relation to each other. They emphasize the central role of modeling, which determines the issues surrounding multidisciplinary design.

In the manufacturing industry, design and development are focused on the product and the product lifecycle. Product lifecycle management (PLM) covers all objects, the lifecycle and the management of changes in solutions, with the aims to reduce the development cycle and its total cost. Concurrent engineering and integrated engineering, corollaries of the introduction of the digital model, are defined. Ontologies, especially product ontologies, are used to develop the semantic level, most often at the level of a trade, one of the current challenges to integrate them into multidisciplinary design.

From the point of view of methods, the lifecycle, explored by mechanical engineering and the manufacturing industry, has also proved to be central in the evolution of the structure. The V-cycle, computer-assisted methods of modeling and simulation allowed for the development of these concepts. The stabilization and generalization of tools, such as STEP, the most widely used standard for the exchange of data throughout the product lifecycle, have equipped these concepts. Finally, the authors illustrate, by showing the ways in which mechanical engineering has moved from research to the establishment of the standard as one of the outcomes, a widely shared agreement followed by a transition into practice that is imposed on everyone. This evolution in the manufacturing industry confirms what is happening in civil engineering through BIM, and the debates underway are coming together, as well as those on systems engineering, ontologies, agile methods, and taking into account the requirements of the operator and/or the user/client.

In Chapter 3, Pierre Benning and Claude Dumoulin deal with the difficult problem of interoperability. To do so, they first recall the reasons and the principles. In a BIM approach, and OpenBIM a fortiori, no software covers all the functions necessary to design and build a structure, and the requirement to use specialized software is inevitable, as is the need to exchange information between software, in order to avoid re-entries, sources of errors, and digital discontinuities. The need for reliable and durable exchanges is therefore immense. The difficulty lies in the fact that this information must remain stable in its formats and properties, both when it is being exported and when it is being imported, regardless of the software.

They introduce the concrete mechanisms of interoperability through standards by clearly showing us how these standards work. The IFC standard is now an ISO standard, based on the STEP file format, and uses object classes, object relationships and object properties – the IFC, a neutral exchange format, prepared and evolving since 1996 by international alliances, first IAI and then bSI. The creation of this ISO standard is an example of the industry's capacity for innovation and organization through international standardization committees. The authors, MINnD actors, who participate in the various standards development committees, give a very synthetic and current overview of development methodologies, for the new classes developed, as well as of the current issues for infrastructures concerning IFCs. They remind us that if the BIM approach requires a description of the world, very precise and operational, in a world of objects, with properties, and relationships between objects, we need to remember that the essence of the collaborative approach of BIM is based on processes and dynamic evolutions.

In Chapter 4, Ana Roxin, Christophe Castaing and Charles-Édouard Tolmer demonstrate how the structuring of information through BIM is guided by the overall objective of building the digital twin of the structure. The information system that is the digital twin, or digital asset, aims at transforming the unstructured data into structured information around which services are developed to allow for its exploitation. The digital twin is based on interoperability, which is itself based on standardization. The digital twin is the dynamic representation of a structure: from its origin, the digital representation must integrate the markers for monitoring its progress within the lifecycle. The first service is data sharing as defined by ISO 19650 (2018). Seeing BIM as the starting point for the creation of the digital twin also implies thinking about the digital twin within its dynamics. Imagining the object to be built in its evolution throughout its lifecycle means understanding the interaction and impacts of the design, construction and operation processes, on the digital twin. In addition, it requires not only modeling the data but also the processes.

The ISO 19650 standard for the digital twin meets two objectives: to represent the asset and to make it operational. It describes two types of requirements: those on the performance of the product and those on the performance of the project processes. On the one hand, it structures the object to be realized, that is, the product or system "to be made", and on the other hand the processes, guaranteeing the lifecycle, that is, the project or system "for making". The use of system engineering and requirements is reinforced by the use of common modeling formalisms.

The digital transition of the construction sector introduces the issue of information valorization into project management. We are moving from file trees, data dictionaries and product catalogs to a graph structured by metadata. New methods from other industries, such as requirements management, systems engineering, knowledge management and PLM (Codohinto and Kiviniemi 2014; Jupp and Singh 2016), are being integrated into BIM and are contributing to its restructuring.

In Chapter 5, Ana Roxin and Christophe Castaing detail two UML-based approaches used for modeling complex systems, such as infrastructure construction projects. The objective of this type of modeling is to make decisions in a complex problem space following different stages of design and implementation processes. The authors situate this approach in the global framework of data modeling and take the ISO 8000 series of standards as a reference framework.

Object-oriented analysis, which came out of object-oriented programming, responded to the need to better manage inheritance. An "object" was then defined as an abstraction in the problem space, keeping the information and properties of this object. An object-oriented analysis then aimed to describe the problem space. A decade later, UML formulated a new definition of the object by putting the focus on aspects of data-processing. A class thus becomes a descriptor for a set of objects sharing the same characteristics. With UML modeling, everything is an object and the analysis aims to be situated in the solution space.

The authors situate the use of ontologies in the evolution of object thinking and the entity-relation approach. Ontologies describe a universe through classes, properties and instances of classes (or individuals). In order to define that an individual belongs to a class, the class must first be described in a formal way by specifying necessary conditions or necessary and sufficient conditions. However, ontologies are also used for the definition of dictionaries and other resources integrated into a BIM project (glossaries and term hierarchies), which do not require a formal language.

In Chapter 6, Lauri Koskela, Saeed Talebi, Algan Tezel and Patricia Tzortzopoulos analyze the recently identified synergy between BIM and Lean processes. A reciprocal perspective on the contributions of BIM to Lean and Lean to BIM is presented in the design, construction and maintenance of structures.

BIM is a collaborative process, consistent with a flexible and economical management for the design of the structure, allowing the choices of concepts to be kept open for the benefit of the performance criteria. It also optimizes the planning of work during the construction phase by facilitating the identification of constraints, reducing waste, errors and conflicts, and optimizing the verification and validation of implementation processes. In the operation and maintenance phases, the benefits of BIM should also enable the implementation of an economical management of the built assets.

Conversely, Lean implies a proactive approach based on rigor, collaboration, experimentation and continuous improvement, which are essential values for the implementation of a mature BIM process. For the authors, the synergies between BIM and Lean processes are constantly evolving and will continue to develop.

In Chapter 7, Rebekka Volk gives an overview of the techniques for generating building data, from scan-to-BIM methods to design-for-deconstruction (DfD) techniques. She describes the advantages, requirements and research perspectives. The centralized and structured management of building information should enable the application of the principles of the circular economy and optimized deconstruction planning. As-built modeling represents a major challenge in optimizing the end-of-life of a building structure.

For already built structures, there are not yet complete, universal, satisfactory commercial BIM-based solutions. The most advanced technologies are able to convert point clouds into digital mock-ups, but the conversion into models suitable for the BIM process is still often inaccurate and difficult to use, especially for complex and irregular structures, such as historic buildings. The BIM deconstruction process suffers from the current lack of standardization.

However, for the author, the use and integration of other digital techniques, related to the exploitation of structured data, its geographical location and components in the BIM deconstruction process, could give rise to an automated and industrialized deconstruction process, more conducive to the circular economy and the optimization of the deconstruction.

In Chapter 8, Hervé Halbout, François Robida and Mojgan A. Jadidi present the complementarities and convergence of BIM and GIS. The foundations and history of each of the shared digital representations are outlined, as well as the evolution of standards, the long-desired convergences that are beginning to take shape, particularly since the agreement between buildingSMART and OGC in 2014. The most widespread formats being CityGML for GIS and IFC for BIM, interoperability is played out, either via intermediary formats, or the compatibility of each of the formats with an RDBMS format. Digital continuity is the objective being sought in a context where data are increasingly acquired by sensors, where the Internet of Things (IoT) links a structure to its territory.

The context of the work for the various institutions is largely described and allows clarification of the evolution of the standards and the formats of exchanges, which are the concretization of these convergences and complementarities between GIS and BIM.

All of these chapters cover a very wide range of topics on BIM, both on the theoretical level and on applications, the originality of the book being to expose the theoretical foundations of modeling and data representations that are used in BIM, while integrating the industrial point of view, which develops not only considerations on the applicability of concepts and conceptual models, but places this approach in perspectives raised by the needs of the industry. We are indeed presenting the results of scientific research on engineering that proposes concepts and methods that can be applied. The reader will therefore find in this book, near complete theoretical contributions and a description of all the issues, representative of those encountered by engineers in the construction sector, who are employing elaborate methods and models in order to implement the BIM approach.

I.3. References

Aish, R. (1986). Building modelling: The key to integrated construction CAD. In *International Symposium on the Use of Computers for Environmental Engineering Related to Buildings*, Tokyo.

Baba, Y. and Nobeoka, K. (1998). Towards knowledge-based product development: The 3D CAD model of knowledge creation. *Research Policy*, 26(6), 643–659.

Bjork, B. (1989). Basic structure of a proposed building product model. *Computer Aided Design*, 21(2), 71–78.

Eastman, C. and Henrion, M. (1977). GLIDE: A language for design information systems. *ACM SIGGRAPH Computer Graphics*, 11(2), 24–33.

Eastman, C. and Siabiris, A. (1995). A generic building product model incorporating building type information. *Automation in Construction*, 3(4), 283–304.

Latiffi, A.A., Brahim, J., Fathi, M.S. (2014). The development of Building Information Modelling (BIM) definition. *Applied Mechanics and Materials*, 567, 625–630.

NBIMS (2015). National BIM Standard – United States Version 3, s.l.: National Institute of Building Sciences buildingSMART Alliance, Washington.

Norme ISO 19650 (2018). Organization and digitization of information about buildings and civil engineering structures, including building information modelling (BIM) – Information management using building information modelling.

Penttilä, H., Rajala, M., Freese, S (2007). Building information modelling of modern historic buildings. Session 13: Modelling – eCAADe 25, 607–613 [Online]. http://papers.cumincad.org/data/works/att/ecaade2007_124.content.pdf.

Van Nederveen, G. and Tolman, F. (1992). Modelling multiple views on buildings. *Automation in Construction*, 1(3), 215–224.

1

Disruptive Technology and Economic Issues

Régine TEULIER[1,2] and Marie BAGIEU[3]
[1]*CNRS, CRG, Institut polytechnique, Paris, France*
[2]*Projet MINnD, Paris, France*
[3]*ESITC Caen, France*

1.1. BIM as a disruptive technology

In order to be able to state that building information modeling (BIM) is a disruptive technology, we will first examine what constitutes a disruptive technology, then analyze how it can be qualified as such, and finally suggest the specific characteristics of BIM that make it a disruptive technology.

A disruptive technology, or disruptive innovation, is a technological innovation that involves a product or service that ultimately replaces a technology that was dominant in the market (Bower and Christensen 1995; Christensen et al. 2015). Disruptive technologies, in contrast to usual technology evolution, impact markets and create many discontinuities, both in client segments, new channel organizations and organizational components. Discontinuous technologies are not always disruptive technologies; they do not necessarily immediately replace the previous service. For example, automobiles did not immediately replace horses. The disruptive technology came about with the introduction of the Model T Ford, which truly transformed the market. Few technologies are intrinsically disruptive; it is often their use that is strategic and that has this effect.

1.1.1. *The concept of disruptive technology*

Disruptive technologies occur differently within the market environment, depending on whether they concern the heart of the firm's activity or its periphery, a product that is at the top or at the bottom of the product hierarchy, a complex or simpler product or service.

Innovation can be introduced into the firm through process innovations, which almost always result in a product innovation. One hypothesis made by some works is that an existing technology can create a disruptive opportunity for a given business activity (Yoon and Lee 2008; Young et al. 2008; Yu and Hang, 2008).

Innovation can be seen as structural (Gatignon et al. 2002) by distinguishing the complexity of the product (the number of its subsystems) or according to the *locus* of innovation in the product hierarchy: core/periphery (Clark 1985; Tushman and Murmann 1998). We can also differentiate the various types of innovation (generational and architectural) or incremental/radical. Finally, we can consider innovation according to the increase in competences versus the destruction of competences, or the acquisition of radically new competences that it implies. The relations between the dimensions of innovation and the processes around competences are particularly important for the evolution of the firm (Tushman and Anderson 1986; Anderson and Tushman 1990). The results of innovation for the firm and sector deserve to be analyzed.

The boundaries between these different concepts are not always clear (Ehrnberg 1995) and as such they may take many forms. The types and characteristics of technological change are varied. The consequences for the firm are reflected in its competences, organization and management. One of the processes in the creation of disruptive innovation in the firm is the central process of destroying/enhancing of competences as described by Tushman and Anderson (1986). They distinguish between two types of innovation: those built on the destruction of skills and those built on their increase: "Competence-enhancing innovation builds upon and reinforces existing competencies, skills and know-how. Competence-destroying innovation obsolesces and overturns existing competencies, skills and know-how" (Gatignon et al. 2002, p. 1107).

The work of Gatignon et al. (2002), analyzing several innovations, has shown that it is more complicated to renew the interaction between competency and organizational devices when it is not a question of destroying competences, but of enhancing competences while introducing new competences from outside the firm (Cohen and Levinthal 1990; Rothaermel 2001). But these are also the most successful innovations.

Some studies provide details related to the difference between old and new knowledge. Katila and Ahuja (2002) distinguish between the depth dimension in firms, which consists of revisiting the firm's traditional knowledge, and the breadth dimension, which is the degree of new knowledge that is explored. Katila and Ahuja have shown that some firms simultaneously engage in both dimensions and that this is likely to maintain a competitive advantage.

Disruptive innovation enters the market in phases that are difficult to decipher at the time they occur. Companies are reluctant to use these innovations because they compete with traditional technologies which work and are profitable. The innovations risk cannibalizing their activities, which are still profitable. At the same time, the firms tend to satisfy the upper end of the customer base and try to improve their margins with the new product (Anderson and Tushman 1990). Initially, the performance is inferior, but disruptive innovation still enters the market (Yu and Hang 2010). Innovation typically involves changes in the business model (Yu and Hang 2010) and difficult adjustments, such as the mix of adopting new and old processes. It particularly serves to compensate for the transient performance gap (Thomke 2003, p. 181). The outcome of innovation is described by Gatignon: "The more complex the innovation, the greater its perceived success, yet the longer it takes to introduce" (2002, p. 1117). In this framework, the firm that adopts a technological innovation early on will be able to maintain a certain lead.

1.1.2. *BIM interpreted as a disruptive technology?*

BIM can be interpreted as a disruptive innovation (Teulier 2017). It is first and foremost a disruptive technology, imposed from the outside by the progression of digitization, as described for example by Yu and Hang

(2008). This primary reason then leads to innovations in design and methods specific to the field. The increase in computer capacity, which has enabled the widespread use of 3D and large databases, has resulted in a technological supply and opportunities, which can be broadly described as digital. This supply, for the major players in the construction industry, met their needs to foresee and better plan, to better control costs and deadlines, the need to have numerous simulations and numerous variants and to better control the product and the margins overall. This combination has been a powerful motivation to adopt these technologies. The complex structure for the world of contracts in the construction sector, between project management, contractors and project owners, was undoubtedly a factor that favored the adoption of renewal through technology.

BIM, as a disruptive innovation in construction, means the use of digital tools and all that this use implies. These implications are, as this book demonstrates, the adaptation of all civil engineering, the redesign of methods and a profound modification of project operations (Boland et al. 2007). As such, BIM is initially difficult to identify as a disruptive technology, since the final physical product does not seem to have changed: what is changed is the project and the way of conducting the project. This only becomes perceptible during the first phase of innovation dissemination. In a second phase where innovation is adopted, it becomes more easily perceptible that it is also the organized digitized data organized in certain formats that become the object of the contract.

Even if the structures built with BIM are similar to those built without BIM, we can consider that the delivered product is not only the physical work, but the work along with a set of digitized data that can be used in tools that have yet to be largely defined for the operation, maintenance and deconstruction/reuse of the construction. This set of digitized data is an opportunity to increase the capacity for action of all the actors who are downstream of the construction delivery, which is the precise definition of "relevance" (Sperber and Wilson 1986). Organized data are highly relevant to all actors since they increase their own capacity for action and flexibility.

1.1.3. *The characteristics of BIM as a disruptive technology*

If we accept that BIM is a disruptive technology, then what are the characteristics that make it one? We can say that BIM touches the heart of the construction system, insofar as it concerns the processing of data: it is the very heart of the company's activity that it impacts, as well as its entire production. The locus of innovation concerns the totality of the interventions made within the framework of construction projects through the processing of all the data. The adoption of a BIM approach is therefore a complex innovation; it affects the entire hierarchy of products and processes.

BIM, like any digital innovation (Yoo et al. 2012), can be considered distributed and combinatorial. It is distributed because it spreads to the periphery of organizations, which increases the heterogeneity of knowledge resources. It requires that other actors in the immediate environment also innovate. It is combinatorial because it allows existing modules to be combined with integrated digital possibilities. It can also be combined with other technological innovations according to the uses and communities, which take hold of them and enrich them with new functions.

Let us look at BIM from a competence perspective. It is a situation where new competences are added to old ones, as described by Gatignon et al. (2002), since civil engineering knowledge remains but must be processed for incorporation into the project. The fact that there is mostly an enhancing competence means greater commercial success for the dominant firms that adopt a BIM approach. However, it also means that since there is no destruction of knowledge, the renewal organizational links and arrangements are a challenge for them.

This may explain why BIM takes time to penetrate companies: it is a complex innovation, which is located at the heart of the business but which also affects all aspects of the construction project. Its type, as a process innovation, means that it has to replace proven routines and processes in the firm that support individual competences and hierarchies. Its characteristics require a significant increase in competences, not only of individuals, but also of teams, departments and the whole organization, which take time to grow, while the old ones are still fulfilling their role in relation to the company's business model.

Several results of the work of Gatignon et al. (2002) can lead to contradictory hypotheses, which we could use for BIM: on the one hand it would spread quickly because it affects the core business of the firm and is strategic; on the other hand, as a complex innovation, its success is perceived as important, though it takes time to be introduced; the fact that BIM relies on new skills, in addition to an increase in competences and that it affects the core of the firm would also argue for rapid generalization. Yoo et al. (2012)

note that combinatory innovations do not just diffuse, they mutate and evolve as they are adopted by heterogeneous communities, which strongly confront their actions. The BIM approach, in a company and in a sector, is therefore built as it is adopted. Finally, BIM would be imposed on the market following a process described by Li et al. (2019), driven by both technological advances as well as the demands and needs of project owners. The product is modified and the client operator has an interest in having digitized data that will be exploitable through appropriate tools.

We propose to summarize below the characteristics that allow us to define BIM as a disruptive technology:

– It is a disruptive technology, which first appeared to be externally "imposed" onto companies in the construction industry, appearing to be more of a process innovation than a product innovation, but which, at least, transforms the product from a physical work to a physical work + data + services.

– The delivery of data must be done in the form of large interoperable and organized data sets, on which services can be grafted.

– BIM is organized around the data and concerns the entire construction project, therefore all the trades and companies that the project brings together.

– It implies cooperation and data sharing between all the companies involved in the project. They can no longer work in "hubs". Cooperation widely becomes an obligation.

– The organization of data and the importance of modeling imply a shift in added value. This change of delivery from a physical product to a physical product + data (and possible services around the data) causes displacement in terms of value creation and competitive advantage.

– The roles and interrelations of all stakeholders are profoundly modified, with engineering and builders having a leading role in the change that BIM represents.

– The role of the owner is emphasized. This role occurs differently for operation and maintenance. Actually, in a transition phase, owners have some difficulty in fully occupying their new role.

– The management of the life cycle of the work is highlighted by the BIM and this recomposition of activities around data sharing (Borrmann et al. 2018).

– Cooperation between the competences of the three kinds of stakeholders is greatly increased: owners, construction project players and software publishers.

– The strong requirement for interoperability implies an increased importance on standardization, in general, and open formats, in particular.

– This technological breakthrough implies a significant enhancement of competences, both in terms of individual and collective, but also organizational and inter-organizational competences.

– BIM is slowly being seen as supporting a global approach and operates as an integrator of other current innovations, such as sustainable and low-carbon development approaches, or future innovations such as artificial intelligence (AI) methods and tools.

1.2. Introduction of BIM in the construction industry: observations from the French construction industry

Two characteristics of the construction industry will determine the adoption and diffusion of BIM as a disruptive innovation:

– Companies are organized by project. In the construction industry, project-based operation has been mastered for a long time and had a very structuring impact on the sector. This project-based organization favors the gradual adoption of technological change and the gradual adjustment of the company's resources (Christensen and Overdorf 2000).

– Companies cooperate on construction sites. Each company obtains the allocation of construction consignments for which, on the construction site, they are obliged to cooperate and exchange data, while keeping to their individual commitments.

Let us consider the three poles for stakeholders that interact to build BIM: the structure owner, the construction project players and the software publishers. By construction project players, we refer to all the actors within the construction industry involved in the project being carried out through a BIM approach. Although each one has its own issues, the BIM tools and approaches that are built depend, above all, on their confrontations. This is illustrated in Figure 1.1.

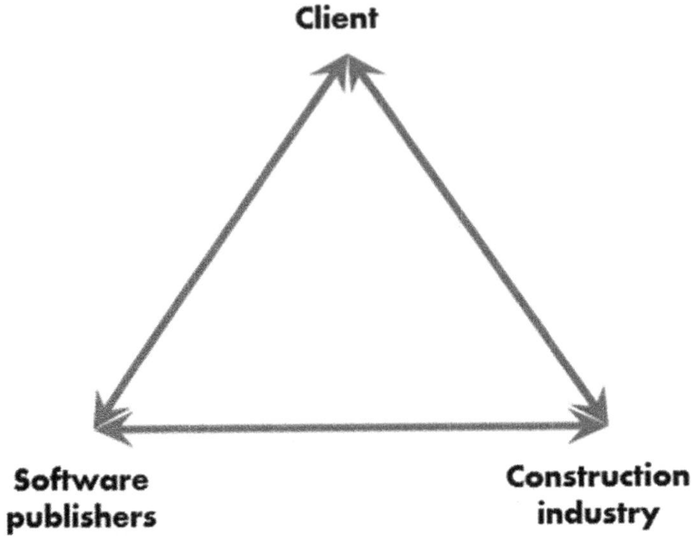

Figure 1.1. *The three main types of actors among which BIM is emerging and disseminating*

For the construction industry, which has to rebuild its business model, the interaction with suppliers, such as software publishers, requires special attention. The engineers of the software publishers already spent long periods of time on-site with their clients, to beta test new versions, install them and migrate customer data into these versions. More globally, their discussion and development work between publishers and construction companies is modified. The development of a range of tools and BIM tool platforms requires integrating the "business vision" of construction companies. The construction project players need software publishers to allow in their solutions the use of neutral formats of the data: for example, IFC, BCF, City GML or others, and allow for co-use with OpenBIM.

For software publishers, this means integrating customer requirements. This on the one hand is a broader and current movement for all companies, but on the other hand calls into question their current business model based on domination by proprietary standards.

The requirements of the owner and of the operator (including maintenance), represent the requirements of the clients for the construction project players. This second type of requirement presents more difficulties to become legible. It is difficult, in particular for the operator, who up until now has not been represented in the building project, to formulate requirements and to follow the design, and then follow it at the construction site, when they themselves are only involved at a distant stage, or even when they are not yet identified, because the operation has not yet been awarded. Project owners, in their calls for tenders, can sometimes have maximalist and contradictory requirements due to a lack of maturity on the subject; nevertheless, they are also the driving force and it is their requirements that push construction firms to accelerate the adoption of BIM approaches. A formulation of customer requirements is also being built, based on the needs of users, such as building managers, who are seeking to procure tools that are useful to them.

With BIM, client requirements have an increased importance in the design and construction processes, but not all obstacles are removed and each of these three poles must renew itself in order to contribute to the co-construction of BIM and to reinvent its business models.

1.2.1. *The digital effect and the transformation of software and platforms*

The use of digital has a very specific role on the design processes, which have been extensively studied (Star and Griesamer 1989; Le Masson et al. 2011; Mougins et al. 2015). In the case of BIM, the disruption introduced by digital is continuous, from design to operations, through construction and maintenance, and to deconstruction and reuse. It is therefore not only the design and its processes that attract the attention of the players in this sector, but the overall restructuring aspect that BIM represents.

In concrete terms, the companies' resource centers validate new software, compatible with the software and know-how already mastered within the company. In general, each project can use these resources at its request. An important point is the great autonomy of construction projects, especially for large projects. The relationship with the different resource centers of the company is therefore often in the form of a cooperative dialogue. This cooperation is complemented by the previously mentioned cooperation with software publishers.

Finally, platforms are a new actor in these transformations. They play a "disruptive" role by interrupting the value chain and can be destructuring or restructuring. They are often built around major digital players like Google, Microsoft and Oracle. Their use reorients the entire life cycle management (Lenfle et al. 2007). The European initiative, Gaia-X, clearly aims to build a data and service platform at a European scale, for all industrial sectors, as well as for public uses. Gaia-X, launched by France and Germany, is supported by the major European digital players (Gaia-X 2020). New platforms specific to the construction industry are emerging (Heiskanen 2021) and address issues within this sector (Kovacic et al. 2020).

What needs to be invented for each project, in addition to a renewed team of engineers, is the package of software chains that link old and new software. We need to reinvent new design methods, but also new ways of exchanging and sharing data with players who were previously peripheral to the central activity of engineering and builders. The articulation of new competence and traditional competence is thus produced in vivo, in projects, and always under cost and time constraints.

1.2.2. *The transformation of all the company's processes*

The project-based structure encourages experimentation and learning by trial and error. The project, which is a component of the construction intervention mode, is particularly suited for two fundamental modes, described by Loch et al. (2006) as factors that reduce uncertainty and risk: experimentation and trial-and-error learning.

It is the project that constitutes the "exploration unit", depending on its specific context, the involvement of the project manager and their team and the possibilities of inter-company cooperation. This is how companies test and adapt. The acquisition of new knowledge is done through exploration projects (Lenfle 2008) and experiments, which mobilize both internal resources, such as the composition of the project team or new hires, and external resources, such as cooperation and co-exploration partnerships (Segrestin 2006; Ben Mahmoud-Jouini 2016). This type of in situ exploration in real projects, carried out within a very constrained framework, offers the opportunity to experiment and to "learn through projects", provided that certain conditions are met: the choice and representativeness of the processes experimented with, the capacity to multiply them and undertake their cost, the strategy put in place to organize and interpret them (Thomke 2003, p. 221).

It is the whole organization of the construction project that is so specific (Ben Mahmoud-Jouini et al. 2004), which changes around the construction to be built. Yoo et al. (2006) showed that the uniqueness and the originality of each construction project, the requirements of the customer, the financial arrangement and the environment of the built construction explain this uniqueness and require a specific organization of the project. The project is constructed at the same time as the building is designed: "In effect, the design of the building and the design of the organization are mutually constitutive – a *bricolage*, to be sure, but a carefully orchestrated one" (Yoo et al. 2006, p. 227).

Businesses have an integrating role in two senses. First, businesses and positions are emerging whose role is to integrate information or activities. For example, the job of the project manager has been profoundly modified, and new roles, such as BIM manager, appear. Second, all professions are being modified since they all have to integrate this understanding of other professions and the concern of the client's needs, which makes it impossible to continue working in "hubs". The concrete specialist and the structural specialist keep to their core activity, theoretical corpus and methods, but with BIM, they can no longer do it separately. Each expert must interact with their own results with the other trades and with the project manager. This is not easy for some experts.

Engineering, as a generic profession, is deeply modified, constituting a revolution in engineering and a cognitive revolution. Modeling is once again central to the engineering profession, more so than calculation (Le Moigne 1977). Modeling is at the heart of BIM and plays out in complex stages. Therefore, the problem is not only the transition from the "as-designed" model to the "as-built" model, but it is necessary to transition to a dynamic model allowing the anticipation of the behaviors of the built structure. For example, in order to help predict what will happen in the middle of the construction's life cycle, even if it is only 30 years away. One difficulty, therefore, is to make the link between the model "as-designed", the model "as-built" and the model of behavior in use or "as-maintained".

Digital continuity, which is not specific to construction, requires the use of common data through a chain of actors, and the repositioning of their roles. It forces us to break down the "hubs"; it is a revolution in the cognitive behaviors and the intellectual services and not just the tools.

Among academic studies, disruptive technology should also be studied within the conceptual framework of works on activity (Van der Berg et al. 2021). Detailed descriptions of activity, such as that by Cristia (2020), which are all exceptionally rare, show, based on the monitoring of a construction site using the BIM approach, that the construction of the BIM model and the entire data environment was as time-consuming as the physical construction itself. Similarly, there is little in-depth work on competences. Gu and London (2010) explain that roles are changed, and new ones are created. Kassem et al. (2018), using the conceptual framework proposed by Succar et al. (2013), and using the four roles of BIM specialists (BIM manager, information manager, BIM coordinator and BIM technician), proposed by Gu and London (2010), propose a detailed profile description for each role. Starting from project-centered competences and organization-centered competences, and crossing competence sets and fundamental competences used by the project, they construct four profiles that list these fundamental competences.

1.2.3. *The management of the project*

Using Morris' works (2013) and ISO 21500: 2012, about project management, we propose this definition for project management: A set of concepts, tools and techniques on how to execute projects on time, within budget, and according to required client specifications in the context of an explicit business strategy.

In the case of the transition phase of BIM approaches, we are still in the control paradigm, which assumes that project management begins once the specifications are defined. We are not yet in a context of permanent innovation, where the needs are poorly defined and are invented as the project progresses with the client. We can therefore make the hypothesis that we are in a controlled type of project management, which is renewed by the recomposition imposed by digital technology.

At the beginning, the innovation in a BIM project lies in the combinatorial (Yoo et al. 2012) or *bricolage* (Yoo et al. 2006) aspect and the way in which new tools are combined with old tools, or old methods, or the way of composing teams that mix new competences with a strong emphasis on integrating digital technology and professional competences focused on the fundamentals. The act of implementing new approaches in real projects leads to tinkering and prototyping channels and adaptations to make it work. These processes have been widely studied for other industry sectors (Henderson 1998; Youman 2011; Ben Mahmoud-Jouini 2016), with Thomke (2003), for his part, having theorized its experimentation.

Project management is impacted by digital technology, particularly by all the processes of collective development and exchange within the project. Therefore, the use of digital tools modifies the exchange of knowledge (Carlile 2002). Cooperative processes require integrating an understanding of the partnership's goals (Zacchary and Robertson 1990), which implies a deep understanding of the other's point of view. The combination of heterogeneous knowledge, belonging to different worlds (Yoo et al. 2012), profoundly renews all modes of collective work and cooperation within the project, as well as its management thereof.

1.2.4. *Project portfolio and corporate strategy*

From the company's position, it is important to capitalize from one project to the next (Loufrani-Fedida and Missonier 2015), and to determine strategies that are transverse to its projects. It is also important to conceive of all the projects as a portfolio that expresses, in a differentiated way, the global technological transformation to be carried out. The strategy cannot only be one of capitalization and bottom-up; it must also be transverse and top-down, to allow for centralized strategic thinking and decision-making. Beyond the feedback and the capitalization of knowledge, it is a question of interpreting a series of experiments carried out in real-world conditions, in order to deduce what can be generalized at the company level. It is therefore not a logic of accumulation that must be implemented, but rather a logic of strategic orientation. This strategic work presupposes an observation of all these experiences, an interpretation and a strategic competency that allows for elaborated orientations as well as the organizational competences to implement them across all levels of the organization (Christensen 1997; Henderson 2006).

In companies in the construction industry, think tanks, often transversal organizations, in addition to projects, also play the role of an "exploration entity". They have, for example, the role of preparing the new thought processes for the company's leaders, carry out internal communications to prepare for change and to facilitate the deployment of new tools, play the role of "promoting the dynamics of renewal" identified among the roles of "exploration entities' by Ben Mahmoud-Jouini (2016, p. 76). However, they are also outward-looking, through their members giving courses in engineering schools and helping to recruit young engineers or participating in inter-company studies and research.

The adoption of a BIM approach is therefore done within the company, through projects, and spillover from precursor projects as Ben Mahmoud-Jouini and Charue-Duboc (2014) have shown for other sectors, through the testing and provision of tools and methods by resource centers. Deployment should not be understood in the strict sense, as in IT. A "BIM approach' is more than just a new generation of software, so adoption processes must be implemented along with the deployment of new tools and innovations. The precursor aspect of the innovation changes the roles in its adoption process (Ben Mahmoud-Jouini and Charue-Duboc 2014).

1.2.5. *Inter-company cooperation*

The BIM approach is of optimal interest if it is adopted by the construction project as a whole, and therefore by all companies working together. However, construction is carried out by allocation of lots, so the adoption of the BIM approach implies the cooperation of inter-company teams within the project. This also means data sharing between teams working with different software chains.

Construction projects, strongly structured by contracts, where cooperation is very generalized and very differentiated according to levels of intervention, make the evolution of practices with strong legal implications very complex. However, the changes must start from the fundamental processes and lead to adaptations of all the systems. Castaing emphasizes, "The BIM, as a digital process, leads to redefine the collaborative relationship even if the responsibilities in the art of building, do not change" (2019, p. 8, *author's translation*).

In the current exploration phase, we can observe that firms are cooperating to design a project that includes exploratory approaches in the use of BIM tools, the choice being made project by project and not just within the firm. Two companies can thus develop in-depth cooperation to implement a BIM approach on a major project, and not do so on another major project that is contemporary with the first. Project leadership is a very important dimension. It is the tangible framework for inter-company cooperation. It is all the more important for the overall success of the project, as activities are increasingly integrated, and as a common data environment (CDE) is imposed on all.

The strategic importance of influencing the evolution of standardization is apparent at all levels. It is a way for the sectors to organize themselves internationally, by preparing for the evolution of standardization. Standardization plays a crucial role, insofar as it is the translation of systems that translate innovation into the procedures and practices of the company, precisely those that will be imposed on it later. It is therefore fundamental, for the sector as well as for the company, to participate in its evolution rather than to suffer its effects.

1.3. Economic issues

The economic challenge for the company implies the redefinition of its activity and its competitiveness around a new business model. This is taking shape around all the processes recomposed around BIM and the new value creation. The company's ability to identify and develop a new approach consisting of the physical public work + data sets adapted to the needs of the operator is decisive, as well as a new way of cooperating with its partners and customers. It is in this context that competitive advantage can occur. It also depends on the company's ability to reorganize its project management methods and its inter-company cooperation methods, and to redefine all of its projects as a portfolio of experimentation projects. That means engaging in a process of experimentation and learning. The appropriation of this disruptive technology by companies presupposes an appropriation by employees, who are accustomed to a different way of working, but also an appropriation by the organization itself, and therefore a change in all the organizational processes, as already mentioned.

The challenge for the sector is to focus on its innovation processes, to control them so as to be active vis-à-vis the other stakeholders. The global control of its processes requires a collective commitment to the task of developing standards, which, by the setting of the standards, strongly influence the processes, or even, like ISO 19650 (2018), standardize the processes themselves. The sector can promote inter-company cooperation to achieve these objectives, through joint research in joint research projects (MINnD S2 2019) or through commitments in learned societies such as buildingSMART International (bSI 2020), through joint actions on the constraints weighing on companies in the sector (redefinition of relations with publishers, joint international action on standards, etc.), through the dissemination of good practice guides, protocols and agreements to be made between companies for contracts, or finally through a global discourse on the cooperation required in the BIM approach in projects.

1.4. Implementation and diffusion of BIM

The adoption of a BIM approach is done through operational projects, according to our observations in the French construction industry. Therefore, operational projects have a certain experimental function. A distinction must be made between the project, the team of a company involved in the multi-company project and the company itself.

The adoption of a BIM approach by a company requires the reinvention of the processes implemented by the company for these projects, as have been shown in the previous paragraphs. They also depend on the CDE that can be adopted at the level of the project itself by the intercompany teams.

OpenBIM, which is currently gaining ground, is accentuating this movement. Based on the application of standards, it can be defined as an implementation of BIM concepts, principles and methods that are operationalized outside of any proprietary format, relying solely on business principles and open, freely shared standards. OpenBIM and the adoption of standards is a way for industries and professional organizations to bring forward the adoption and diffusion of BIM in their sector.

OpenBIM is a robust ecosystem for the development of a BIM approach, over the entire lifecycle of the works. It is not comparable to any type of project and company, whether for design, construction or operation of the constructions. OpenBIM is defined by a model-based approach (e.g. using IFC), which organizes data and describes the digital asset, and a process-based approach (e.g. using ISO 19650, 2018), which follows standardized contract processes. The implementation of OpenBIM was enabled in 2020 because this widely shared work production standard resulted from the industry players themselves and reached a level of maturity and accumulated tangible results in international standards committees. IFC 4.3 and ISO 19650 Part 1 and Part 2 (2018) were available for tender since 2020.

Succar and Kassem (2015) examined the conceptual structures of BIM adoption. They distinguish between BIM implementation, which is the successful adoption of BIM tools and a BIM approach within an organization, as well as BIM diffusion, which is the rate of penetration of the tools and data exchanges across the market. BIM implementation is seen as occurring in three phases: adoption readiness, performance capability and performance maturity.

Types of diffusion models depend on types of influence for diffusion: either vertical (e.g. state influence) or collateral relationships (such as professional organizations).

Succar and Kassem (2015) and Kassem and Succar (2017), through two articles, try to build an understanding of the dynamics of BIM adoption, in the form of a survey (which excludes France). From these results, they build a conceptual framework and then a proposal for types of BIM adoption models. In this work, diffusion is seen from a Schumpeterian[1] perspective,

---

1 Schumpeter (1883–1950) proposed an economic theory (in 1911) linking innovation to the creation of a company and to the specific figure of the entrepreneur, in a fundamental one-off event in the evolution of capitalism: the creation of a company. He then proposed the concept of innovation by clusters after a major innovation (1939). This is an economist vision, which has been modified through the managerial approaches of recent decades. Now the figure of the project manager, design expertise, multidisciplinary approaches, the introduction of customer and user points of view, marketing methods, and concurrent engineering, are renewing the ways of considering innovation. Currently, innovation, its conditions, processes, and consequences are the focus of numerous studies in various disciplines. Innovation is thus seen as a central process in the knowledge economy. The new conditions of innovation are often studied in the context of new technologies and, for example, the emergence and role of start-ups.

which has been relativized in theoretical works on innovation for the past decade. Currently, innovation is no longer seen as a consequence of R&D work, systematic and sequentially organized.

1.5. Measuring BIM maturity

Maturity models (MM) have been available since the late 1970s. Crosby's model, the QMMG or Quality Management Maturity Grid (Crosby 1979), the first identified, originated in the field of quality management. This type of model is referred to as a descriptive maturity grid or matrix (Fraser et al. 2002): it contains textual descriptions for each activity at each maturity level. The Software Engineering Institute, funded by the US Department of Defense, developed the Capability Maturity Model (CMM) in the 1990s (Paulk et al. 1993). It takes a different approach to the maturity matrix, looking at key performance indicators (KPIs), which can be realized as the maturity level increases. Fraser et al. (2002) identify a third MM, based on a questionnaire and a Likert-type rating scale (Likert 1932). This last model allows for evaluation of the integration maturity of a process in a quantified way.

The maturity of companies with regard to BIM has long been a topic of concern and much work has been published to assess this maturity, particularly by proposing a maturity structure by levels. Since the publication of the first edition of the NBIMS standard (NBIMS 2007), several BIM maturity assessment tools have indeed been proposed, the precursors being, in particular, the NBIMS BIM Capability MM (BIMCMM) (NBIMS 2007), the iBIM maturity model (Bew et al. 2008) and the BIM maturity matrix (Succar 2009; Succar et al. 2012). These different models are the subject of numerous comparative studies (Giel and Issa 2014; NBIMS 2015; Wu et al. 2017; Ferraz et al. 2020).

The UK maturity model, also known as the iBIM or BIM Wedge model, was developed by Bew and Richards (2008). There are several versions of the basic model with subtle but significant differences; the one presented in Figure 1.2 appeared in the UK Government Construction Client Group (GCCG) report in 2011.

The iBIM model identifies specific capability levels covering technology, standards, guides, classifications and delivery.

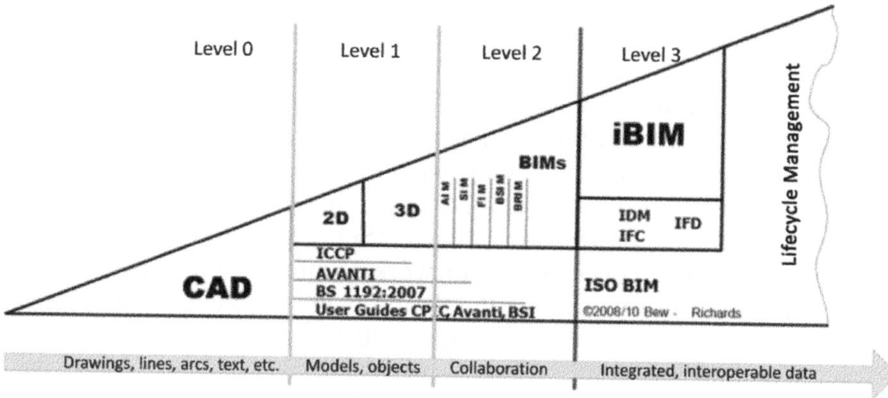

Figure 1.2. *BIM maturity levels (GCCG 2011)*

These levels show the progress in the company's adoption of BIM:

– Level 0: 2D CAD, unstructured, generally on paper.

– Level 1: mix of 2D CAD and 3D digital mock-up. The data are structured and the process integrates a collaboration tool.

– Level 2: 3D environment managed in separate BIM tools, specific to each collaborator (architect, design offices, builders), with structured data attached and possible exchanges of digital models. Level 2 allows the use of 4D and 5D BIM (4D BIM model with added cost data).

– Level 3: this is the most mature level. It corresponds to iBIM or integrated BIM. The integration of processes and data is complete. It is close to a concurrent engineering process. The principle is that a single model is stored on a centralized server, accessible by all participants and throughout the life of the work via formats such as IFC, CityGML, BCF, methods and methodological tools and structuring information.

Maturity levels are now understood by the standards of the exchanges and by the access to the heterogeneous environment of the databases. Maturity levels can be seen in their relationship to standards. BS 1192 (2007), replaced by the ISO 19650 (2018) series of standards, is the first element that defines Level 3. It corresponds to the exchange of 3D data and implies that this exchange is framed by the standards.

The ISO 19650 standard proposes stages that correspond to the technology stages, so the 2016 maturity levels are now captured by the stages of ISO 19650. ISO 19650's (2018) statement "Organization and digitization of information about buildings and civil engineering works, including building information modelling (BIM) – Information management using building information modelling" was published in 2018. It is made up of two parts: Part 1 on concepts and principles and Part 2 on data and echoes for the design and construction phases. ISO 19650 refocuses the BIM approach to a broader conceptual framework of information management, to build both the physical work and the digital twin. Theoretical contributions also go in this direction (Succar and Poirier 2020).

At the 3BIM level, we no longer aggregate models; we federate models and this is only possible using norms and standards. It is a model-based approach, not a service-based approach. BIM 4D allows for construction simulations, and BIM 5D allows for cost simulations. When approaching maturity levels, we start by distinguishing between an analytical model and a static model. The static model is the model as-designed, and the analytical model as-built also called a dynamic model, because it can generate simulations. This is what allows us to make a leap, a disruptive in terms of the management of the project and the management of the work.

Part 1 of ISO 19650 specifies a set of definitions for the concepts that are proposed, as summarized in Table 1.1.

Part 2 of ISO 19650 describes a set of processes for exchanging information and deliverables in the design and construction cycle, including tasks, roles and responsibilities.

The report by Northumbria University, on behalf of the Center for Digital Built Britain (CDBB), in partnership with the UK BIM Alliance (Kassem et al. 2020), presents a study of the different BIM maturity tools, with respect to ISO 19650. It links the elements assessed in each project BIM maturity tool to the corresponding clauses of ISO 19650, with the objective of understanding the relevance of the assessment offered by a BIM maturity tool against the ISO standard and the extent of coverage for a standardized approach to information management.

This cross-analysis allowed the authors to identify commonalities and differences between the analyzed tools and to make recommendations for BIM maturity assessment, which must be encouraged to help organizations make the change. This assessment is currently ill-equipped. Skills should play a greater role in the bidding, consultation and execution phases. The adoption of the ISO 19650 standard should go in this direction. The authors of the Northhumbria report (Kassem and Li 2020) recommend the development of a multi-level framework to do this, as well as the development of a method built on the 19650 standard and centered on collaborative work.

Concept	Definition
Building information modeling	Use of a shared digital representation of a built asset to facilitate the design, construction and operation processes and to form a reliable basis for decision-making.
BIM execution plan	Plan that explains how the information management aspects of the consultation will be carried out by the implementation team.
Level of information required	A framework that defines the scope and granularity of the information.
Information	A named and persistent set of retrievable information within a file, system or application storage hierarchy. Example: subdirectory, information file (including template, document, table, and calendar) or a distinct subset of an information file such as a chapter or section, a layer or a symbol.
Common data environment (CDE)[2] BS 1192:2007	An agreed-upon source of information about a project or asset, used to collect, manage and disseminate each piece of information through a managed process.
Requirements management	To specify the information that members of the asset or project's supply chain must provide as part of their work and to inform the owner of the asset/organization/project.

Table 1.1. *The key concepts of the ISO 19650 standard*

2 These items are recognized by ISO 19650, without being core concepts. They are defended as key concepts by the EFCA (2018).

1.6. Conclusion

BIM can be characterized as a technological breakthrough for the construction industry. This characterization makes it possible to review the work accrued in other fields and to propose criteria for the analysis and observation of an ongoing phenomenon. BIM is not just a chain of software and a shared common data platform; all the uses and processes around it are modified. The product delivered, itself resulting from the construction project, is different. It is no longer just a physical structure but also access to shared data responding to the uses of operation, maintenance and deconstruction/reuse, what can broadly be called the digital twin.

In general, the technological breakthrough only fully plays its disruptive role and is fully visible when it arrives on the market and recomposes it. It therefore takes quite a long time to observe. In addition, BIM affecting all processes, essentially the heart of the company, itself takes a long time to be introduced. As a digital innovation, it can appear both in a distributed and combinatorial way, with certain modules developed in certain parts of the projects recombining with existing modules.

As with all technological breakthroughs, the link between disruption and knowledge is central. BIM is essentially a process of enhancing skills: old skills are still essential, but they need to be augmented with knowledge of digital tools. Finally, shared data, at the center of all exchanges, put cooperative processes at the center of shared modeling and imply that business visions open up to each other.

Assessing maturity for BIM adoption can always be interesting. The ISO 19650 standard has profoundly renewed this assessment; it replaces the model in levels by stages (internships) from which methods could be proposed.

As a technological breakthrough, BIM creates a new technological, cognitive, social and organizational environment and can be the support for other innovations and transformations of uses (IA, low carbon guidelines, etc.). Initial signs show that companies are gradually adopting BIM and are preparing to use it as a support for low-carbon initiatives.

1.7. References

Anderson, P. and Tushman, M.L. (1990). Technological discontinuities, dominant designs: A cyclical model of technological change. *Administrative Science Quaterly*, 35, 604–633.

Ben Mahmoud-Jouini, S. (2016). Le numérique au service des entités dédiées à l'innovation de rupture. *Revue française de gestion*, 254(76), 65–84.

Ben Mahmoud-Jouini, S. and Charue-Duboc, F. (2014). Le déploiement d'innovations inter-filiales au sein d'une multinationale. *Management International*, 18, 41–58.

Ben Mahmoud-Jouini, S., Midler, C., Garel, G. (2004). Time-to-market vs. time-to-delivery managing speed in engineering, procurement and construction projects. *International Journal of Project Management*, 22, 359–367.

Bew, M., Underwood, J., Storer, G. (2008). Going BIM in a commercial world. Sophia Antipolis, eWork and eBusiness. In *Architecture, Engineering and Construction*, Zarli, A. and Scherer, R. (eds). CRC Press, Boca Raton.

Boland, R.J., Lyytinen, K., Yoo, Y. (2007). Wakes of innovation in project netconstructions: The case of digital 3D representations in architecture, engineering, and construction. *Organization Science*, 18(4), 631–647.

Borrmann, A., König, M., Koch, C., Beetz, J. (eds) (2018). *Building Information Modeling*. Springer International, Springer, Heidelberg, London, New York.

Bower, J. and Christensen, C.M. (1995). Disruptive innovation: Catching the wave [Online]. Available at: https://hbr.org/1995/01/disruptive-technologies-catching-the-wave [Accessed on 2 February 2023].

BS 1192 (2007). BUILDING SMART ALLIANCE 1192 Building Information Management. A Standard Framework and Guide to BS 1192 [Online]. Available at: https://www.cpic.org.uk/publications/a-standard-framework-guide-to-bs-1192/ [Accessed 2 February 2023].

buildingSMART (2020). buildingSMART [Online]. Available at: https://www.buildingSMART.org/ [Accessed 25 January 2021].

Carlile, P.R. (2002). A pragmatic view of knowledge and boundaries: Boundary objects in new product development. *Organization Science*, 13(4), 442–455.

Castaing, C. (2019). Le BIM ou le paradoxe français. In *De la recherche à l'industrialisation*, Tolmer, C.E. and Teulier, R. (eds). Eyrolles, Paris.

Christensen, C.M. (1997). *The Innovator's Dilemma. When New Technologies Cause Great Firms to Fail*. Harvard Business School Press, Boston.

Christensen, C.M. and Overdorf, M. (2000). Meeting the challenge of disruptive change. *Harvard Business Review*, 78, 66–76.

Christensen, C.M., Raynor, M., MacDonald, R. (2015). What is disruptive innovation? [Online]. Available at: https://hbr.org/2015/12/what-is-disruptive-innovation [Accessed 2 February 2023].

Clark, K. (1985). The interaction of design hierarchies and market concepts on technological evolution. *Research Policy*, 14, 235–251.

Codinhoto, R. and Kiviniemi, A. (2014). BIM for FM: A case support for business life cycle. In *Product Lifecycle Management for a Global Market: 11th IFIP WG 5.1 International Conference*, PLM 2014, Yokohama, Japan, July 7–9, 2014, Revised Selected Papers 11. Springer, Berlin, Heidelberg.

Cohen, W. and Levinthal, D. (1990). Absorptive capacity: A new perspective on learning and innovation. *Administrative Science Quaterly*, 35, 128–152.

Cristia, E. (2020). Architectural project and BIM digital model. An ethnographic investigation realised on a famous architectural project. PhD Thesis, CNAM, Paris.

Crosby, P.B. (1979). *Quality Is Free: The Art of Making Quality Certain*. McGraw-Hill, New York.

EFCA (2018). BIM and ISO 19650 from a project management perspective. Information management using building Information modelling [Online]. Available at: https://www.efcanet.org/publications/bim-booklet [Accessed 20 November 2021].

Ehrnberg, E. (1995). On the definition and measurement of technological discontinuities. *Technovation*, 15(7 September), 437–452.

Ferraz, C., Rocha Loures, E., Deschamps, F. (2020). BIM maturity models evaluated by design principles [Online]. Available at: https://easychair.org/publications/preprint_open/cfHw [Accessed 25 January 2021].

Fraser, P., Moultrie, J., Gregory, M. (2002). The use of maturity models: Grids as a tool in assessing product development capability. In *IEEE International Engineering Management Conference*, Cambridge.

GAIA-X (2020). GAIA-X [Online]. Available at: https://www.datainfrastructure.eu/GAIAX/Navigation/EN/Home/home.html [Accessed 15 October 2021].

Gatignon, H., Tushman, M.L., Smith, W., Anderson, P. (2002). A structural approach to assessing innovation construct development of innovation locus: Type, and characteristics. *Management Science*, 48(9), 1103–1122.

Giel, B. and Issa, R. (2014). Framework for evaluating the BIM skills of building owners. *Computing in Civil and Building Engineering*, 552–559.

Gu, N. and London, K. (2010). Understanding and facilitating BIM adoption in the AEC industry. *Automation and Construction*, 19(8), 988–999.

Heiskanen, A. (2021). Openbuilt platform accelerates digitalization of the construction-industry [Online]. Available at: https://aec-business.com/openbuilt-platform-accelerates-digitalization-of-the-construction-industry/ [Accessed 27 October 2020].

Henderson, K. (1998). The role of material objects in the design process. A comparison of two design cultures and how they contend with automation. *Sci. Techn. Human Values*, 23(2), 139–173.

Henderson, K. (2006) The innovator's dilemma as a problem of organizational competences. *The Journal of Product Innovation Management*, 23, 5–11.

ISO 19650 (2018). Organization and digitization of information about buildings and civil engineering works, including building information modelling (BIM) – Information management using building information modelling [Online]. Available at: www.iso.org/fr/standard/68078.html.

ISO 21500 (2012). Guidance on project management [Online]. Available at: https://www.iso.org/standard/50003.html [Accessed 2 February 2023].

Jupp, J.R. and Singh, V. (2016). A PLM perspective of BIM research initiatives. *International Journal of Product Lifecycle Management*, 9(3), 180–197.

Kassem, M. and Succar, B. (2017). Macro BIM adoption: Comparative market analysis. *Automation in Construction*, 81, 67–98.

Kassem, M., Raoff, N.L., Ouahrani, D. (2018). Identifying and analyzing BIM specialist roles using a competency-based approach. In *Creative Construction Conference*, Ljubljana.

Kassem, M., Li, J., Kumar, B., Malleson, A., Gibbs, D.J., Kelly, G., Watson, R. (2020). Building information modelling: Evaluating tools for maturity and benefits measurement. Centre for Digital Built Britain: London, UK. [Online]. Available at: https://www.researchgate.net/profile/Jennifer-Li-4/publication/344819191_Building_Information_Modelling_Evaluating_Tools_for_Maturity_and_Benefits_Measurement/links/609191f0a6fdccaebd091af1/Building-Information-Modelling-Evaluating-Tools-for-Maturity-and-Benefits-Measurement.pdf [Accessed 2 February 2023]

Katila, R. and Ahuja, G. (2002). Something old, something new: A longitudinal study of search behavior and new product introduction. *Academy of Management Journal*, 45(6), 1183–1194.

Kovacic, I., Honic, M., Sreckovic, M. (2020). Digital platform for circular economy in AEC industry. *Engineering Project Organization Journal*, 9(1), 1–16.

Le Masson, P., Hatchuel, A., Weil, B. (2011). The interplay between creativity issues and design theories: A new perspective for design management studies? *Creativity and Innovation Management*, 20(4), 217–237.

Le Moigne J.-L. (1977). *La Théorie du système général. Théorie de la modélisation.* Presses Universitaires de France, Paris.

Lenfle, S. (2008). Exploration and project management. *International Journal of Project Management*, 26(5), 469–78.

Lenfle, S., Ben Mahmoud-Jouini, S., Derousseaux, C. (2007). New product development in a platform-driven organization: Towards platform lifecycle management. Academy of Management, Philadelphia.

Li, J., Kassem, M., Ciribini, A.L.C., Bolpagni, M. (2019). A proposed approach integrating DLT, BIM, IOT and smart contracts: Demonstration using a simulated installation task. International Conference on Smart Infrastructure and Construction, Cambridge.

Likert, R. (1932). A technique for the measurement of attitudes. *Archives of Psychology*, 22(140), 55.

Loch, C., DeMeyer, A., Pich, M. (2006). *Managing the Unknown: A New Approach to Managing High Uncertainty and Risk in Projects*. Wiley, Hoboken.

Loufrani-Fedida, S. and Missonier, S. (2015). The project manager cannot be a hero anymore! Understanding critical skills in project-based organizations from a multilevel approach. *International Journal of Project Management*, 33, 1220–1235.

MINnD (2019). MINnD [Online]. Available at: https://www.minnd.fr/ [Accessed 25 January 2021].

Morris, P.W.G. (2013). *Reconstructing Management*. Wiley-Blackwell, New York.

Mougin, J., Boujut, J.F., Pourroy, F., Poussier, G. (2015). Modelling knowledge transfer: A knowledge dynamics perspective. *Concurrent Engineering Research and Applications*, 1(12).

NBIMS (2007). National BIM Standard – United States Version 1, s.l.: National Institute of Building Sciences buildingSMART alliance.

NBIMS (2015). National BIM Standard – United States Version 3, s.l.: National Institute of Building Sciences buildingSMART alliance.

Norme ISO 19650 (2018). Organisation et numérisation des informations relatives aux bâtiments et ouvrages de génie civil, y compris modélisation des informations de la construction (BIM) – Gestion de l'information par la modélisation des informations de la construction.

Paulk, M., Curtis, B., Chrissis, M., Weber, C. (1993). Capability maturity model, version 1.1, s.l. *IEEE Software*, 10(4), 18–27.

Rothaermel, F.T. (2001). Incumbent's advantage through exploiting complementary assets via interfirm cooperation. *Strategic Management Journal*, 22(6/7), 687–699.

Segrestin, B. (2006). Toward a new form of governance for inter-firm cooperation? Lessons from the Renault-Nissan Alliance. *International Journal of Automative Technology and Management*, 6(2), 199–213.

Sperber, D. and Wilson, D. (1986). *Relevance: Communication and Cognition*. Blackwell's, Oxford.

Star, S.L and Griesemer, J.R. (1989). Institutional ecology, translations and boundary objects: Amateurs and professionals in Berkeley's Museum of Vertebrate Zoology. *Social Studies of Science*, 19(3), 387–420.

Succar, B. (2009). Building information modelling framework: A research and delivery foundation for industry stakeholders. *Automation in Construction*, 18(3), 357–375.

Succar, B. and Kassem, M. (2015). Macro-BIM adoption: Conceptual structures. *Automation in Construction*, 57, 64–79.

Succar, B. and Poirier, E. (2020). Lifecycle information transformation and exchange for delivering and managing digital and physical assets. *Automation and Construction*, 112, 103090.

Succar, B., Sher, W., Williams, A. (2012). Measuring BIM performance: Five metrics. *Architectural Engineering and Design Management*, 8(2), 120–142.

Succar, B., Sher, W., Williams, A. (2013). An integrated approach to BIM competency acquisition, assessment and application. *Automation in Construction*, 135, 174–189.

Teulier, R. (2017). BIM, the technological breaking for civil engineering? Toward a synergy between industry and academic. *IJ3DIM*, 6(3), 6.

Thomke, S.H. (2003). *Experimentation Matters: Unlocking the Potential of New Technologies for Innovation*. Harvard Business Press, Boston.

Tushman, M.L. and Anderson, P. (1986). Technological discontinuities and organizational environments. *Administrative Science Quarterly*, 31(3), 439–465.

Tushman, M.L. and Murmann, J.P. (1998). Dominant designs, technological cycles and organizational outcomes. *Research in Organizational Behavior*, 20, 231–266.

Van den Berg, M., Voordijk, H., Adriaanse, A. (2021). BIM uses for deconstruction: An activity-theoretical perspective on reorganising end-of-life practices. *Construction Management and Economics*, 39, 323–333.

Wu, C., Xu, B., Mao, C.X.L. (2017). Overview of BIM maturity measurement tools. *Journal of Information Technology in Construction (ITcon)*, 22, 34–62.

Yoo, Y., Boland, R.J., Lyytinen, K. (2006). From organization design to organization designing. *Organization Science*, 17(2), 215–229.

Yoo, Y., Boland, R.J., Lyytinen, K., Majchrzak, A. (2012). Organizing for innovation in the digitized world. *Organization Science*, 23(5), 1398–1408.

Yoon, B. and Lee, S. (2008). Patent analysis for technology forecasting: Sector-specific applications. In *IEEE Engineering Management Conference, IEMC Europe*, Detroit.

Youman, R.J. (2011). The effects of physical prototyping and group work on the reduction of design fixation. *Design Studies*, 52, 115–138.

Young, G.K., Jong, H.S., Sang, C.P. (2008). Visualization of patent analysis for emerging technology. *Expert Systems with Applications*, 34, 1804–1812.

Yu, D. and Hang, C.C. (2008). Creating candidate technologies for disruptive innovation: A case study approach. In *Proceedings of IEEE ICMIT 2008*, Bangkok.

Yu, D. and Hang, C.C. (2010). A reflective review of disruptive innovation theory. *International Journal of Management Review*, 12, 435–452.

Zachary, W. and Robertson, S.P. (1990). Introduction. In *Cognition, Computing, and Cooperation*, Robertson, S.P., Zachary, W., Black, J.B. (eds). Ablex Publishing, Norwood.

2

3D Engineering and Lifecycle Management of Manufactured Products

Benoît EYNARD[1,2], Alexandre DURUPT[1], Matthieu BRICOGNE[1,3] and Julien LE DUIGOU[1]

[1]*Roberval, Université technologique de Compiègne, France*

[2]*Inetum, Meudon, France*

[3]*Groupement francophone des utilisateurs de Catia, Paris, France*

2.1. Introduction

Since the 2000s, 3D modeling and product lifecycle management (PLM) have profoundly changed the way systems and manufactured products are designed, industrialized and mass-produced (Le Duigou et al. 2011; Eynard et al. 2014). This chapter re-examines some of the key elements that structure today's so-called "integrated design" approaches and attempts to offer perspectives for the transfer and adoption of best practices, from the manufacturing industry to the construction and building sector.

This chapter successively addresses the concepts of digital mock-up, product lifecycle management, the role of product modeling and associated standards, collaborative and multidisciplinary design and certain developments in practices related to the deployment of systems engineering, as well as the beginnings of the application of "agility" principles to project management.

Building Information Modeling,
coordinated by Régine TEULIER and Marie BAGIEU.

2.2. Digital mock-up

2.2.1. *How to define a digital mock-up*

Proposed in the 1990s by the European AIT – DMU BP (Advanced Information Technology in Design and Manufacturing – Digital Mock-Up Business Process) project consortium, a digital mock-up is defined as follows: "an extended digital representation of the product used as a platform for product/process development, communication and validation during all phases of the product life" (translated by author).

The digital mock-up includes computer-aided design (CAD) files or 3D models but is not limited to them (Sadoul 2000). It integrates a structured and hierarchical representation of all engineering documentation for the design, industrialization, production, use and dismantling of manufactured products. Therefore, 3D modeling cannot even be considered the main component of a digital mock-up.

The basic element of a digital mock-up is, in our opinion, the bill of materials. This is an ordered list of all the components and parts that make up the product. Generally represented in a tree-like form, the nodes are the components, and the branches are the hierarchical "parent-child" type relationships between the components. Commonly, this bill of material (BOM) is structured (classified) in such a way that it represents the theoretical assembly (termed: "as-assembled") of the product. This classification is a legacy of the CAD "skeleton" modeling approach, also called "top-down", where the principle is to organize the modeling tree with a first node corresponding to a set of geometric elements for spatial positioning of components, such as component positioning points. This set is called the "skeleton" because the geometric elements that comprise it can give it this appearance. Each component modeled in CAD is then linked by assembly constraints to this set. Therefore, updating and modification are simplified because the assembly constraints refer to this single part. This avoids "assembly loops", which often result in editing problems when using larger CAD models. The skeleton approach, due to its organization, also allows for the distribution of models between experts as quickly as possible. Indeed, the simplification of updating and modifying allows for the simultaneous and remote modeling of several parts of the product, even if they are in a contact or kinematic relationship (pivot, ball joint, etc.). Through this logic, the skeleton represents a basic contract between the principal actors and defines the interfaces and bounding boxes, allowing for the exchange of information and better collaboration. It should be noted that the transposition of this concept to the construction sector does not seem to have been the subject of work at this stage. The fact that the issue of strong spatial integration, that is, by achieving a very high degree of compactness of the building's technical systems, does not seem to be a major concern for the sector; this may explain the current lack of interest.

This concept of tree modeling is found in the computerized structuring of the bill of materials for a digital model. For example, in technical data management systems (SGDT or PDM – product data management, Eynard et al. 2004), this nomenclature is known as engineering bill of material (E-BOM), which will generally be transformed into manufacturing bill of material (M-BOM) during the product industrialization stage, as shown in Figure 2.1.

These different interrelated nomenclatures make it possible to structure the collaboration and management of technical data or product data, in particular through the use of "views", "configurations" and "version management systems", presented in the following section.

2.2.2. *Views, configurations and versions of a digital mock-up*

The need for collaboration and management of technical data comes from the fact that engineering tasks around the product are parallelized, the objective being to reduce the time of development and industrialization of products. Concurrent engineering, defined as "an engineering method that consists of involving all the actors of a project, from the very beginning, with the understanding of the objectives and the set of activities that will have to be carried out", became a standard practice in the manufacturing industry in the mid-1990s (Sohlenius 1992, translated by author).

Integrated engineering, defined as "an approach that allows for the integrated and simultaneous design of products and related processes, including production and support, [and] is intended to enable designers to take into account, from the outset, all phases of the product life cycle from conception to retirement, including quality, costs, deadlines and user requirements" (NF X 50-415 1994, author's translation) is also widely used in the industry.

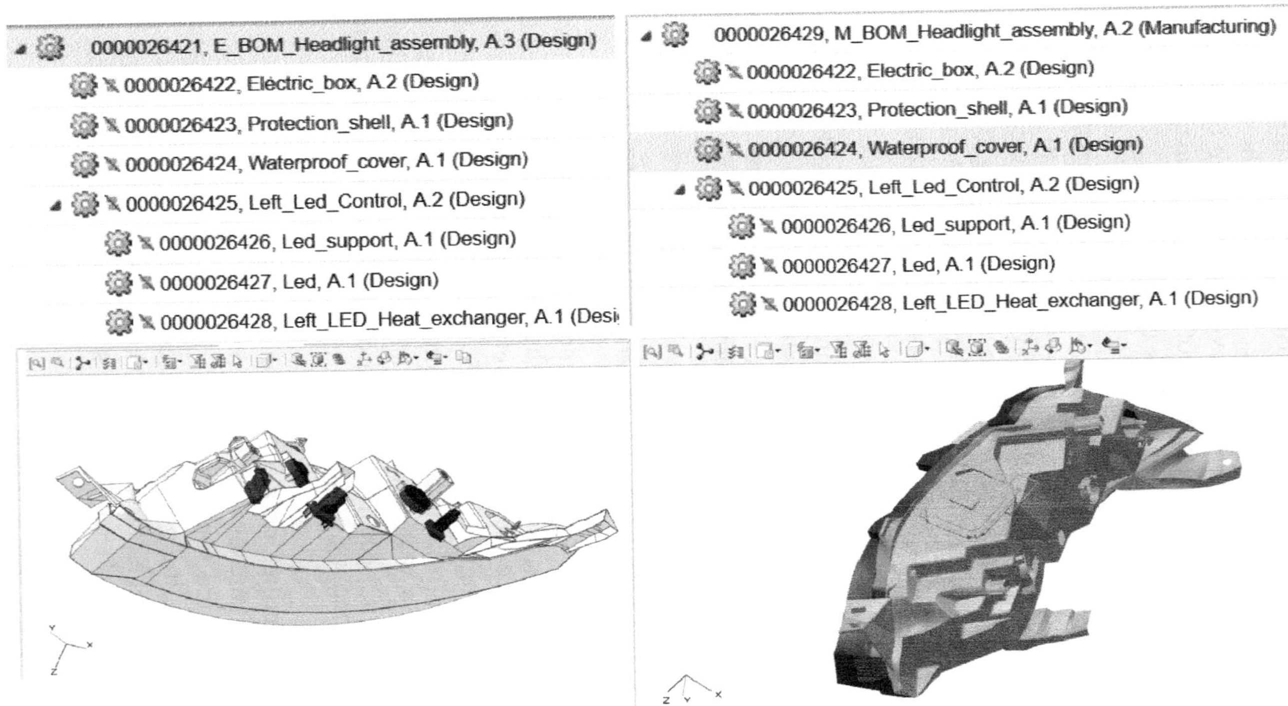

Figure 2.1. *Transformation from the E-BOM to the M-BOM; work realized at the UTC by the students of the UV module TN21 of the UTC using PTC' s Windchill PDMLink 11 solution*

In the digital mock-up, nodes are components, usually called "items", to which engineering data (CAD models, documents, text, reports, etc.) are attached. However, in order to make the best use of this digital model, different mechanisms exist, allowing each expert to realize a content adapted to their specific needs. Among these mechanisms, we can note the notions of:

– Views: this is about organizing the tree structure and the relevant engineering data according to an expertise. The example of the organization of the adaptation of the BOM according to the phases of the product life cycle is a good example. This BOM "as-designed" (E-BOM) can thus be transformed into a BOM "as-manufactured" (M-BOM).

– Configurations: this is the organization of the E-BOM according to the product's variants and options. For example, for a car, the following configurations are as follows: three-door, five-door, convertible, station wagon, etc. Of course, not all options are compatible with each other, and this is where the issue of product diversity management comes in. For the unconfigured product, that is, with all the available options, we speak of a 150% bill of materials.

– Versions: this is the traceability and backup of the digital mock-up at different times or at different stages (milestones) set by the development project.

These different concepts and associated functionalities facilitate collaboration. Beyond this collaborative aspect, the digital model is an important element because it structures PLM. This is presented in the following section.

2.3. Integration of the product lifecycle

2.3.1. *Lifecycle management*

These different methods and concepts of information management continue to evolve and to be combined, giving rise to new developments in design methods and numerous hybrid methods. We consider this to be an evolution, a *continuum* of design methods over time that is explained by several factors:

– the complexity of products is constantly increasing (more technologies, more functions, more businesses to integrate, etc.);

– the servicing of products and the arrival of the concept of product/service systems;

– the customization of products and the ever-increasing demands of customers;

– the evolution of tools (the arrival of digital technology, 3D modeling, the Internet, etc.);

– the increasing consideration toward each stage of the product life cycle.

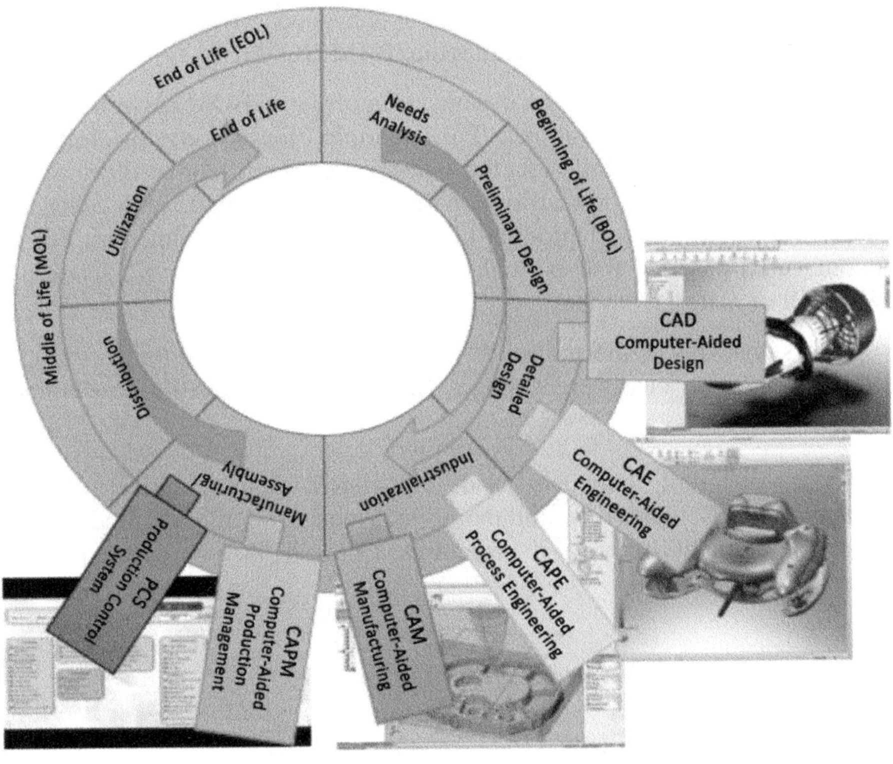

Figure 2.2. *PLM system and business applications*

Taking this last factor into account gave rise in the 2000s to a new paradigm, PLM, and its evolution a few years later with the advent of connected objects, closed-loop PLM (see section 2.3.2). PLM can be defined as a business strategy that aims to create, manage and share all the information about the definition, manufacture, maintenance and recycling of a manufactured product, throughout its life cycle, from the preliminary design to the end of its life (Amann 2002).

PLM is usually associated with a set of applications related to product development processes such as XAO (computer-aided X tasks), PDM, enterprise resources planning (ERP) (Figure 2.2, Terzi et al. 2010).

PLM issues affect all objects (product, process, resources and organization), the entire life cycle (requirements engineering, preliminary design, detailed design, production engineering, manufacturing/assembly, distribution, use/maintenance, end-of-life) and the generation, optimization and management of solution changes. PLM is mainly aimed at improving system quality, reducing the development cycle and total cost. It also aims to reduce the environmental impact of the system and to improve its societal impact.

2.3.2. *Closed-loop lifecycle management*

Behind this very broad definition of PLM, there are nevertheless a number of limitations. It is very difficult to collect information about the product when it is manufactured and even more so once it is distributed. The information coming from the middle of life (MOL) or the end of life (EOL) of the product is very much underexploited in current PLM systems. Based on this observation, the concept of closed-loop PLM was proposed to collect and process MOL and EOL product information, using the new possibilities offered by connected objects, communication technologies and the Internet of Things. The concept of closed-loop PLM was developed during the European FP6 project PRO-MISE and can be defined as follows: a closed-loop PLM system allowing all actors who play a role during the life cycle of a product (project managers, designers, manufacturers, maintenance operators, recyclers, etc.) to track, manage and control product information at any phase of the life cycle (design, manufacturing, MOL and EOL), at any time and from any place in the world (Kiritsis 2011).

The objective is to allow information flows to go beyond the customer, to allow designers to have feedback on the use of the product, its maintenance, or its end of life. Figure 2.3 shows the main information flows of closed-loop PLM as detailed in Jun et al. (2007).

The product sends information about its status and environment to the PLM agent. Each agent collects all the information about the product in real time, for each stage of its life cycle, and at each site (production, distribution, use, etc.). They send all this information to the PLM system, which aggregates it and then gives access to this information to any individual or organization with the appropriate rights.

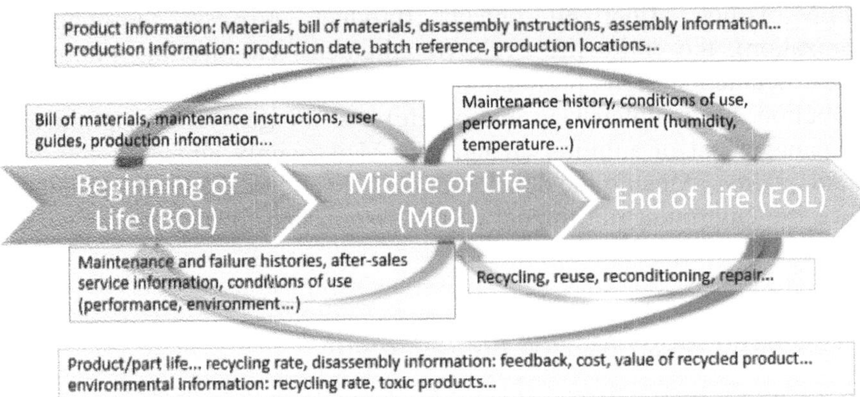

Figure 2.3. *Core information flows of the closed-loop PLM*

Up until now, product optimizations for the MOL and EOL phases have been quite difficult due to the lack of accurate and relevant information. By integrating closed-loop concepts into PLM, it becomes possible to collect MOL information on usage patterns, product location, environment, failures and life expectancy. All this information can be very useful toward improving the design of an existing product as well as a new product. Integrating EOL information is also very promising; it is possible, for example, to estimate the volume of product at the end of its life in a given geographical area, but also to know the quality of the recycled material in order to define the most adequate recycling process and maximizing the value of the recycling. The information from MOL and EOL is therefore available for designers with the objective of improving their present and future designs (Le Duigou 2017). This information is also very useful for all other stakeholders in the life of the product (distributor, repairer, recycling channel, etc.).

This structuring of the phases of the product life cycle seems to be transposable to the construction sector, although there are many important differences. Today, building information modeling is still very much BOL oriented, meaning that the digital model is mainly designed for the act of building the structure. The adaptation or redesign of this model to facilitate the MOL or EOL stages is already a topical issue, even if the heterogeneity actors' allegiances make the task difficult and advocate in favor of a systematic transition through standardization. For example, while the automotive industry has every interest in integrating sensors into its vehicles and setting up data collection and processing platforms to facilitate vehicle maintenance, especially if the vehicle is used as part of a service offered to the user (Mahut et al. 2016), manufacturers are not necessarily encouraged to build a structure aimed at facilitating its operation. The delivery of the digital mock-up and the delivery of the complete documentation about the engineered structure in digital format still seems to be more of a legal obligation than a real reason for upgrading to digital.

Application interoperability and database integration are the main issues from a digital technology perspective. A PLM approach across the entire product lifecycle, from requirements engineering to service retirement, requires reliable, complete and efficient data models and the exploration of an effective digital continuity. We will study product models and meta-models that date back to the 1990s alongside the appearance of the first PDMs and integrated engineering, and then we will discuss the product ontologies that began appearing within our mechanical engineering community in the 2010s, and which are now widely employed in scientific works.

2.4. Models, standards and product ontologies

2.4.1. *Models and product standards*

Product data models are used to structure a large amount of information necessary for the design of a product. Product data models were introduced by Kjellberg and Schmekel (1992) and Krause et al. (1993) at the beginning of the 1990s, and then developed by several authors, including Bernard (2000) and Roucoules and Tichkiewitch (2000), with the aim of managing all the information relating to the product in a multi-actor and multi-view context. All these models manipulate information concerning the product and the associated processes, with the objective of structuring the information in order to facilitate its exchange and reuse. We will review the product process organization (PPO) model resulting from national research as well as the STEP (Standard for the Exchange for Product model data, ISO 10303) and the core product model (CPM) that was the result of international works.

PPO was developed in the RTNL IPPOP project (Integration of Product Process Organization for engineering Performance improvement: Robin et al. 2007; Noël and Roucoules 2008). It proposes a product model allowing to formalize the knowledge about the product (function, structure, behavior) and to link it to expert tools (e.g., CAD), a process model that allows for tracing and capitalizing on the evolutions of the knowledge, as well as an organization model for facilitating multi-objective decision-making.

Initially developed at National Institute of Standards and Technology (NIST), CPM proposes a generic product model based on artifacts (components of the product) aggregating the three views of function, form and behavior (Fenves et al. 2008). It has been completed with several extensions:

– open assembly model, detailing the geometric definition, the kinematic behavior of the product and all the tolerance data;

– master product model and design-analysis integration model, resulting from a desire to link geometric models to idealized simulation models;

– product family evolution model, allowing us to take into account the product evolutions. This extension ensures traceability between the different versions and improvements made to the product.

STEP is the most widely used standard for the exchange of data over the entire product life cycle (Rachuri et al. 2008). Developed under the responsibility of ISO, within the TC184/SC4 subcommittee on “Industrial Data”, this standard specifies, among other things, several application protocols (APs). These APs define several specialized data models for one or more domains. Examples include AP239 (Product Life Cycle Support, PLCS, ISO 10303-239 2012), AP242 (Managed Model-Based 3D Engineering, ISO 10303-242 2020), AP238 (Model-Based Integrated Manufacturing, ISO 10303-238 2020), AP233 (Systems Engineering, ISO 10303-233 2012) or AP209 (Multi-Disciplinary Analysis and Design, ISO 10303-209 2010). It should be noted that AP233 and AP239 are modular and extensible according to the application, whereas the more specialized APs such as AP209, AP242, or AP238 are not modular and extensible.

These models are based on the combination of a function/structure/ behavior view, which according to Gero (1990) allows for a complete understanding of a system, and a product/process/resource view, which defines the three different types of business objects used in a PLM approach. These product models can be seen as light ontologies (without adding reasoning capabilities), which prefigured product ontologies that have appeared in the scientific community since.

2.4.2. *Product ontologies*

Ontologies define a domain by identifying the different concepts that compose it, and the links between them, their properties as well as axioms and rules about them. According to a review of the bibliography by El Kadiri and Kiritsis (2015), the main roles of ontologies in the context of integrated design and PLM are to serve:

– a common and verified source of knowledge that is used and shared by the actors (human or software) of the product life cycle;

– a database;

– knowledge base;

– a bridge between different domains;

– a mediator to ensure interoperability;

– contextual search facilitator;

– Linked Data[1] facilitator.

Among the best-known product ontologies in the field of PLM and integrated design, we can cite the following examples:

– OntoPDM, a product ontology based on ISO 10303 and IEC 62264 managing heterogeneous data, such as those describing materials used in the product, relationships between components and products, versions, manufacturing tools, etc. (Panetto et al. 2012);

1 Linked Data (Bizer et al. 2009) aims to promote the publication of structured data on the Web, not in the form of isolated data hubs, but by linking them together to form a global network of information using Web standards such as HTTP and URI.

– OntoSTEP (Barbau et al. 2012), an ontology based on the CPM, OAM data models as well as the models specified by AP203, 214, and 239 of ISO 10303 STEP. These models have been combined and enriched with semantic and reasoning capabilities. OntoSTEP thus covers geometric data, data related to functions, requirements, behaviors and product design choices.

Ontologies meet a number of needs concerning the management of data, both information and knowledge, particularly at the semantic level. These tools and concepts are one of the major development paths for structuring, sharing and exchanging knowledge in a language that can be used by humans and interpreted by machines. Nevertheless, two obstacles appear concerning their optimal use within the framework of PLM:

– it is necessary to be able to provide ontologies, which answer a requirement as broadly as possible, in order to ensure their genericity, but which is not already covered by other ontologies;

– many design domains are not yet structured in terms of knowledge, or at least not in a language that can be interpreted by a machine.

While these models and ontologies offer possibilities for structuring all product-related information, in practice they are too often still attached to a given discipline. One of the current challenges for the manufacturing industry is therefore to promote new practices for the design of multidisciplinary products, that is, those resulting from the integration of contributions from different know-how and business expertise.

2.5. Multidisciplinary design

Designed systems tend to be more and more integrated. This integration is moving in the direction of functional integration, that is, the aggregation of the maximum number of functions in a single system, but also in the direction of spatial or physical integration, mentioned at the beginning of the chapter, which aims to reduce the size and weight of the system (Warniez et al. 2012). These different types of integration, relative to the product, are also the source of additional technical complexity when compared to a less integrated product. This complexity is generally not the sum of the technical complexities of each of the trades, because phenomena of interdependence of information appear, considerably increasing the complexity of the design. These interdependencies, sometimes also called couplings, exist both at the level of data and expertise (multi-domain interdependencies) and at the level of the various physics involved – multi-physics couplings (Bricogne et al. 2016). These different integration phenomena have been highlighted and remain important issues in the design of so-called multidisciplinary products (Zheng et al. 2016). The two most well-known examples are mechatronic products and cyber-physical systems (CPSs) (Hehenberger et al. 2016).

Mechatronics has been defined by the NF E01-010 standard as "a process aiming at the synergic integration of mechanics, electronics, automation and computer science in the design and manufacture of a product in order to increase and/or optimize its functionality. The objective of mechatronics is to maximize the added value comparing to the simple sum of the added values of functions taken separately" (2008, author's translation). This definition clearly shows the challenge represented by this application framework. It is not a question of creating more or less automated systems, but rather of taking advantage of the integration of various fields involved in the design of such systems (Schöner 2004).

CPSs refer to the generation of systems that require tight integration of computing, communication, and control technologies to achieve stability, performance, reliability, robustness and efficiency in the management of physical systems in many application domains (Rajkumar et al. 2010). Even if the context of CPS development brings out specific needs in this domain, a large majority of the integration issues encountered are similar to those identified in the design of systems involving multidisciplinary teams, such as mechatronics.

In the same way, systems from the construction sector can be considered multidisciplinary systems, requiring the collaboration of business experts with their own know-how, semantics and tools. We then generally observe two categories of problems in the context of the multidisciplinary design of manufactured products. The first type of problem is related to the organization of the design, while the second is related to the management of products' technical data (Abramovici and Bellalouna 2007). The problems related to the organization of the design are linked to coordination and the synchronization of different disciplines issues, the latter having specific approaches and processes that are decomposed into specific activities, tasks and renderings (Merlo and Girard 2004). The problems induced by technical design data management are related to the fact that the editing tools and data management methods for the different disciplines are heterogeneous (Lefèvre et al. 2014). Different interoperability techniques exist, such as the product ontologies presented earlier, but remain, to date, poorly deployed. In the following paragraphs, we present two main families of design concepts and techniques which, after analysis of the bibliography and industrial practices, appear relevant to the construction sector.

2.6. Systems engineering

As illustrated in Figure 2.4, the study of work related to multidisciplinary product design concepts and techniques can be structured into four levels, named approaches, processes, methods and tools for product development (Guérineau et al. 2018). Precise definitions of these different levels are available in the referenced paper. This four-level model also allows for a better understanding of the impact that the deployment of a concept or technique can be expected to have on the strategic, tactical and operational vision of a company (Kéradec 2012). The analysis of industrial practices and scientific publications related to mechatronics and CPSs reveals that large families, large sets of concepts and techniques, emerge and fundamentally structure the way manufacturers currently practice multi-disciplinarity. Among these sets, we introduce systems engineering and agility, because both of them seem relevant for the design of multidisciplinary systems, such as mechatronic systems or CPSs, mentioned earlier.

Systems engineering (SE) is "a general methodological approach that encompasses the set of activities appropriate to design, evolve, and verify a system that provides a cost-effective and efficient solution to a customer's needs while satisfying all stakeholders" (AFIS 2012, translated by author). It is a recognized approach to support multidisciplinary product development and is typically associated with the product development process model called the V-cycle (Dieterle 2005; Kleiner and Kramer 2013). This process is very general and begins with the identification of user requirements and ends with user validation. It is broken down into two main phases: the so-called top-down phase of product decomposition and definition and the so-called bottom-up phase of integration and recomposition (Figure 2.5, US Department of Transportation 2007). Other processes, such as the W-model (Barbieri et al. 2014), enrich, complete and clarify the V-cycle.

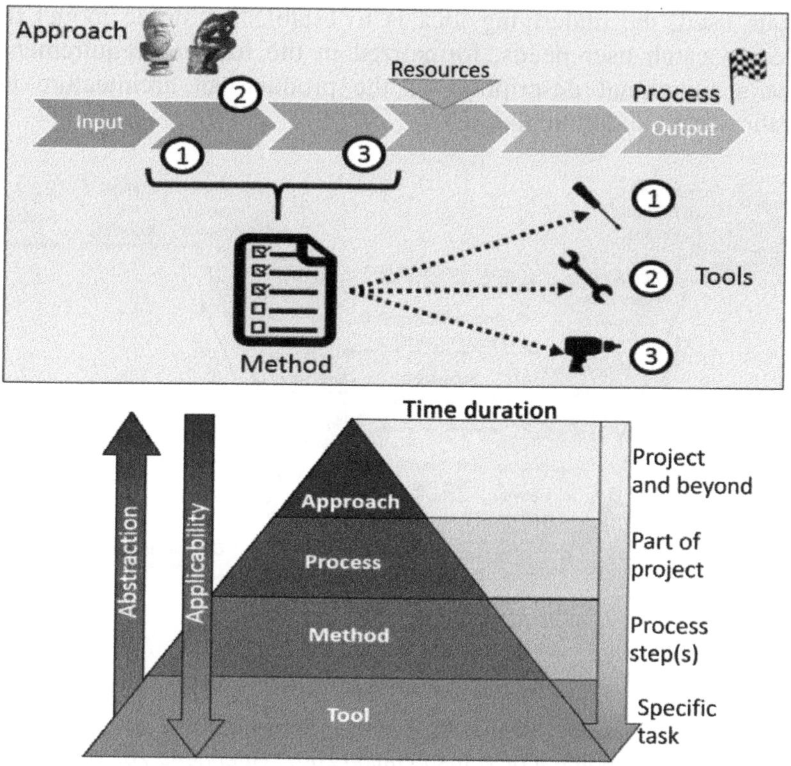

Figure 2.4. *Structuring concepts and techniques used to design multidisciplinary systems into four levels: approaches, processes, methods and tools (Guérineau 2021)*

Based on these processes, many references encourage multidisciplinary product designers to use model-based methods and model-based engineering (Dieterle 2005; Kleiner and Kramer 2013; Couturier et al. 2014; Kernschmidt et al. 2018). This illustrates the clear interest of the scientific communities in these methods. Indeed, models are a recognized way to master the increasing complexity of products (Bricogne 2015). Different authors also explore a number of modeling techniques, such as process-oriented, aspect-oriented, object-oriented or functional modeling. Also related to modeling techniques, some authors promote modeling languages such as UML or SysML, or system simulation languages (or 0D/1D simulation), such as *modelica* (Cao et al. 2011). Whatever the modeling technique used, the underlying idea is to establish a single model that is intended to catch user needs, formalized in the form of requirements, to propose a functional description of the product, an architecture and a verification and validation plan.

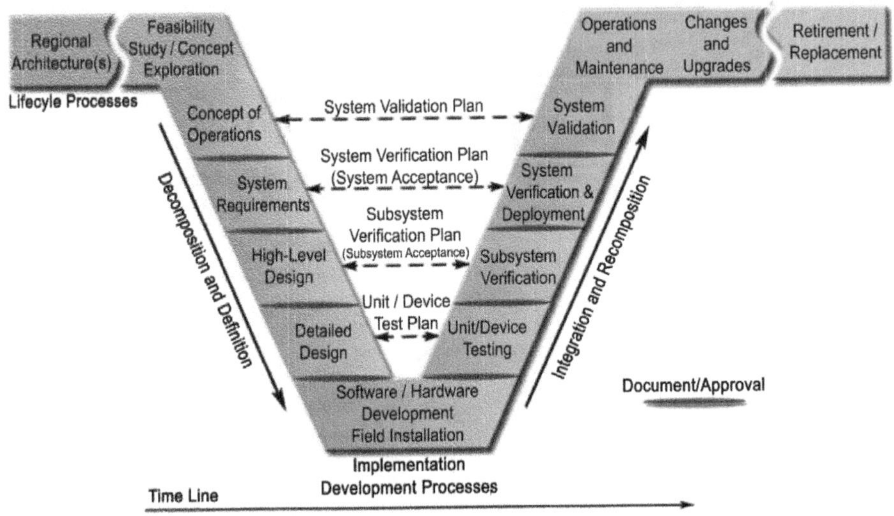

Figure 2.5. *The V-cycle model (Source: based on US Department of Transportation 2007; Guérineau 2021)*

2.7. Agility and digital transformation: the contribution of new collaboration processes

The second set of concepts and design techniques which seems to be adapted to all types of multidisciplinary products, is the so-called "agile" set. If the founding principles of the agile approach were introduced in the *Agile Manifesto*[2], these upshots are numerous, ranging from Scrum processes – certainly the best known to date – to extreme programming, via design thinking. These processes are based on tools, the best known of which are user stories, sprints and backlogs. Different works aim at transposing these concepts and techniques (Bricogne et al. 2014; Bricogne 2015; Goevert and Lindemann 2018) while others aim at hybridizing these methods with other approaches (Stelzmann 2012; Mabrouk et al. 2018).

2 Available at: www.agilemanifesto.org.

Further to hybridization, we are now witnessing the emergence of new practices, resulting from the appropriation and adaptation by various practitioners, of approaches such as lean product development (LPD), user experience design (UXD) and Scrum (Gill 2015). These new practices reveal that actors who used to work on the design, industrialization and production of manufactured systems, that is, predominantly hardware, are now reappropriating the concepts of agility to propose new collaboration processes, adapted to their specificities. They are thus moving beyond the generally highly codified practices of "turnkey" methods to overcome the generally observed limitations of agile methods: uniqueness of the team and its relatively small size, co-located and monodisciplinary team.

Some of the main principles are then adapted according to the companies:

– systematic collaboration with the client or their representative;

– transparency and trust in the team, within the company and even with external collaborators;

– light process: reduction of contractualization phases in favor of productivity and operational functions;

– valuing individuals and their interactions more than processes and tools;

– adapting to change, even late in the process, rather than conforming to initial plans.

Combining this type of approach while "securing the project", based in particular on systems engineering, remains a challenge today, and some research work is already focusing on this type of challenge. Today, the automotive, aeronautics and microelectronics[3] sectors are widely considering this promising combination, which could also appeal to the construction industry during the next decade.

2.8. References

Abramovici, M. and Bellalouna, F. (2007). Integration and complexity management within the mechatronics product development. In *Advances in Life Cycle Engineering for Sustainable Manufacturing Businesses*, Umeda, Y. and Takata, S. (eds). Springer, London.

---

3 Available at: www.afis2020.utc.fr.

AFIS (2012). *Découvrir et comprendre l'ingénierie système*. Cépaduès, Toulouse.

Amann, K. (2002). *Product Lifecycle Management: Empowering the Future of Business*. CIM Data, Inc., Ann Arbor.

Barbau, R., Krima, S., Rachuri, S., Narayanan, A., Fiorentini, X., Foufou, S., Sriram, R.D. (2012). OntoSTEP: Enriching product model data using ontologies. *Computer-Aided Design*, 44(6), 575–590.

Barbieri, G., Fantuzzi, C., Borsari, R. (2014). A model-based design methodology for the development of mechatronic systems. *Mechatronics*, 24(7), 833–843.

Bernard, A. (2000). Modèles et approches pour la conception et la production intégrée. *Journal européen des systèmes automatisés*, 34(2/3).

Bizer, C., Heath, T., Berners-Lee, T. (2009). Linked data – The story so far. In *Semantic Services, Interoperability and Web Applications: Emerging Concepts*, Sheth, A. (ed.). IGI Global, Hershey.

Bricogne, M. (2015). Méthode agile pour la conception collaborative multidisciplinaire de systèmes intégrés : application à la mécatronique. PhD Thesis, Université de Technologie de Compiègne and Ecole de Technologie Supérieure de Montréal.

Bricogne, M., Rivest, L., Troussier, N., Eynard, B. (2014). Concurrent versioning principles for collaboration: Towards PLM for hardware and software data management. *International Journal of Product Lifecycle Management*, 7(1), 17–37.

Bricogne, M., Le Duigou, J., Eynard, B. (2016). Design processes of mechatronic systems. In *Mechatronic Futures – Challenges and Solutions for Mechatronic Systems and their Designers*, Hehenberger, P. and Bradley, D. (eds). Springer, London.

Cao, Y., Liu, Y., Paredis, C.J.J. (2011). System-level model integration of design and simulation for mechatronic systems based on SysML. *Mechatronics*, 21(6), 1063–1075.

Couturier, P., Lô, M., Imoussaten, A., Chapurlat, V., Montmain, J. (2014). Tracking the consequences of design decisions in mechatronic systems engineering. *Mechatronics*, 24(7), 763–774.

Dieterle, W. (2005). Mechatronic systems: Automotive applications and modern design methodologies. *Annual Reviews in Control*, 29(2), 273–277.

El Kadiri, S. and Kiritsis, D. (2015). Ontologies in the context of product lifecycle management: State of the art literature review. *International Journal of Production Research*, 53(18), 5657–5668.

Eynard, B., Gallet, T., Nowak, P., Roucoules, L. (2004). UML based specifications of PDM product structure and workflow. *Computers in Industry*, 55(3), 301–316.

Eynard, B., Le Duigou, J., Bricogne, M. (2014). Des SGDT au product lifecycle management. *Documentaliste : sciences de l'information*, 51(1), 34–37.

Fenves, S., Foufou, S., Bock, C., Sriram, R.D. (2008). CPM2: A core model for product data. *Journal of Computing and Information Science in Engineering*, 8(1), 1–14.

Gero, J.S. (1990). Design prototypes: A knowledge representation schema for design. *AI Magazine*, 11(4), 26–36.

Gill, A.Q. (2015). Learning enterprise agile software engineering: Scaling agility at the enterprise level. In *Proceedings of the 24th Australian Software Engineering Conference (ASWEC)*. Adelaide.

Goevert, K. and Lindemann, U. (2018). Further development of an agile technique toolbox for mechatronic product development. In *Proceedings of the 15th International Design Conference – DESIGN*. Dubrovnik.

Guérineau, J. (2021). Le développement de produits multidisciplinaires à l'ère du numérique et de la connectivité ubiquitaire. PhD Thesis, Université de Technologie de Compiègne and École de Technologie Supérieure de Montréal.

Guérineau, J., Rivest, L., Bricogne, M., Durupt, A., Eynard, B. (2018). Towards a design-method selection framework for multidisciplinary product development. In *Proceedings of the 15th International Design Conference – DESIGN*. Dubrovnik.

Hehenberger, P., Vogel-Heuser, B., Bradley, D., Eynard, B., Tomiyama, T., Achiche, S. (2016). Design, modelling, simulation and integration of cyber physical systems: Methods and applications. *Computers in Industry*, 82, 273–289.

ISO 10303-209 (2010). Application protocol: Multidisciplinary analysis and design. International Organization for Standardization, Geneva.

ISO 10303-233 (2012). Application protocol: Systems engineering. International Organization for Standardization, Geneva.

ISO 10303-238 (2020). Application protocol: Model-based integrated manufacturing. International Organization for Standardization, Geneva.

ISO 10303-239 (2012). Application protocol: Product life cycle support. International Organization for Standardization, Geneva.

ISO 10303-242 (2020). Application protocol: Managed model-based 3D engineering. International Organization for Standardization, Geneva.

Jun, H.-B., Kiritsis, D., Xirouchakis, P. (2007). Research issues on closed-loop PLM. *Computers in Industry*, 58 (8/9), 855–868.

Kéradec, H. (2012). Epistémologie et didactique de la gestion : le cas du concept de décision. PhD Thesis, Conservatoire National des Arts et Métiers, Paris.

Kernschmidt, K., Feldmann, S., Vogel-Heuser, B. (2018). A model-based framework for increasing the interdisciplinary design of mechatronic production systems. *Journal of Engineering Design*, 29(11), 617–643.

Kiritsis, D. (2011). Closed-loop PLM for intelligent products in the era of the Internet of Things. *Computer-Aided Design*, 43(5), 479–501.

Kjellberg, T. and Schmekel, H. (1992). Product modelling and "information-integrated" engineering systems. *CIRP Annals*, 41(1), 201–204.

Kleiner, S. and Kramer, C. (2013). Model based design with systems engineering based on RFLP using V6. In *Proceedings of the 23rd CIRP Design Conference*, Bochum.

Krause, F.L., Kimura, F., Kjellberg, T., Lu, S.Y., Alting, L., Elmaraghy, H.A., Week, M. (1993). Product modeling. *CIRP Annals*, 42(2), 695–706.

Le Duigou, J., Bernard, A., Perry, N., Delplace, J.-C. (2011). Generic PLM system for SMEs: Application to an equipment manufacturer. *International Journal of Product Lifecycle Management*, 6(1), 51–64.

Lefèvre, J., Charles, S., Bosch-Mauchand, M., Eynard, B., Padiolleau, E. (2014). Multidisciplinary modelling and simulation for mechatronic design. *Journal of Design Research*, 12(1/2), 127–144.

Mabrouk, A., Penas, O., Plateaux, R., Barkallah, M., Choley, J.-Y., Akrout, A. (2018). Integration of agility in a MBSE methodology for multidisciplinary systems design. In *Proceedings of the IEEE International Systems Engineering Symposium – ISSE*, Rome.

Mahut, F., Bricogne, M., Daaboul, J., Eynard, B. (2016). Product-service systems for servitization of the automotive industry: A literature review. *International Journal of Production Research*, 55(7), 2102–2120.

Merlo, C. and Girard, P. (2004). Information system modelling for engineering design coordination. *Computers in Industry*, 55(3), 317–334.

NF E01-010 (2008). Mécatronique – Vocabulaire. AFNOR, La Plaine Saint-Denis.

NF X50-415 (1994). Ingénierie intégrée : concepts généraux et introduction aux méthodes d'application. AFNOR, La Plaine Saint-Denis.

Noël, F. and Roucoules, L. (2008). The PPO design model with respect to digital enterprise technologies among product life cycle. *International Journal of Computer Integrated Manufacturing*, 21(2), 139–145.

Panetto, H., Dassisti, M., Tursi, A. (2012). ONTO-PDM: Product-driven ontology for product data management interoperability within manufacturing process environment. *Advanced Engineering Informatics*, 26(2), 334–348.

Rachuri, S., Subrahmanian, S., Bouras, A., Fenves, S., Foufou, S., Sriram, R. (2008). Information sharing and exchange in the context of product lifecycle management: Role of standards. *Computer-Aided Design*, 40(7), 789–800.

Rajkumar, R., Lee, I., Sha, L., Stankovic, J. (2010). Cyber-physical systems: The next computing revolution. In *Proceedings of the 47th ASME Design Automation Conference – DAC*, New York.

Robin, V., Rose, B., Girard, P. (2007). Modelling collaborative knowledge to support engineering design project manager. *Computers in Industry*, 58(2), 188–198.

Roucoules, L. and Tichkiewitch, S. (2000) CoDE: A cooperative design environment: A new generation of CAD systems. *Concurrent Engineering Research and Applications*, 8(4), 263–280.

Sadoul, D. (2000). La maquette numérique : un exemple de développement industriel. *Mécanique et industries*, 1(4), 373–382.

Schöner, H.-P. (2004). Automotive mechatronics. *Control Engineering Practice*, 12(11), 1343–1351.

Sohlenius, G. (1992). Concurrent engineering. *CIRP Annals*, 41(2), 645–655.

Stelzmann, E. (2012). Contextualizing agile systems engineering. *IEEE Aerospace and Electronic Systems Magazine*, 27(5), 17–22.

Terzi, S., Bouras, A., Dutta, D., Garetti, M., Kiritsis, D. (2010). Product lifecycle management – From its history to its new role. *International Journal of Product Lifecycle Management*, 4(4), 360–389.

US Department of Transportation (2007). Systems engineering for intelligent transportation systems. US Department of Transportation, Washington.

Warniez, A., Penas, O., Soriano, T. (2012). About metrics for integrated mechatronic system design. In *Proceedings of the 9th France-Japan & 7th Europe-Asia Congress on Mechatronics (MECATRONICS)/13th International Workshop on Research and Education in Mechatronics (REM)*, Paris.

Zheng, C., Le Duigou, J., Bricogne, M., Eynard, B. (2016). Multidisciplinary interface model for design of mechatronic systems. *Computers in Industry*, 76(24), 24–37.

3

Interoperability Through Standards: IFC, Concepts and Methods

Pierre BENNING[1,2] and Claude DUMOULIN[1,2]
[1] *Bouygues – Travaux Publics, Paris, France*
[2] *Projet MINnD, Paris, France*

3.1. Introduction

Exchange formats to describe the information associated with an infrastructure are necessary in order to ensure interoperability between the many software applications of the actors involved and to guarantee the continuity of the data throughout the life of the infrastructure. A standardization of these formats meets these requirements.

The development of standards requires the contribution of two types of experts: those from professions linked to the infrastructures defining their information exchange requirements, and exchange format professionals. This is illustrated through the principles, concepts and methods used for the development of Industry Foundation Classes (IFC) object classes describing the exchanged information.

Interoperability is illustrated by the development of IFC classes for infrastructure, coordinated by buildingSMART International, including the alignments (IFC-Alignment), structures (IFC-Bridge), tunnels (IFC-Tunnel), roads (IFC-Road) and rail lines (IFC-Rail).

Building Information Modeling,
coordinated by Régine TEULIER and Marie BAGIEU. © ISTE Ltd 2023.

3.2. OpenBIM and interoperability

Building information modeling (BIM) can be defined as "the use of a shared digital representation of a built asset to facilitate design, construction, and operation processes, and form a reliable basis for decision making" (Norm ISO 19650-1, 2018, translated by author).

The main contributions of BIM (Dumoulin 2018) are as follows:

– a triple geometric organization: spatial, by function or system responding to a function, by business;

– analytical organizations interfacing with simulations or optimizations;

– model consistency control functionalities (interfaces between business lines and insertions within its context);

– navigation in the project via its business components;

– a separation between the storage of information and its visualization (non-visual information, such as a calculation note or a contract, is attached to the objects of the model);

– an aid for decision-making;

– traceability of modifications and decisions.

OpenBIM is a collaborative approach applicable to the entire lifecycle, based on open standards and work processes. Combined with other standards such as PLCS (Project Life Cycle Support) from ISO 10303-239 (2012), OpenBIM could even be the basis for project modeling and work modeling (Tarandi 2011).

3.2.1. *The requirements for exchanges*

For any construction project, during its entire lifecycle, information exchanges between actors are numerous and frequent, given the volume of data to be generated, processed and validated.

Defining the ownership rights of each piece of data is therefore fundamental to tracing the person responsible for an item of information.

Finally, it is essential to be able to manage modifications and their repercussions on the associated information, as each piece of data must be considered in its surrounding context, which constrains it.

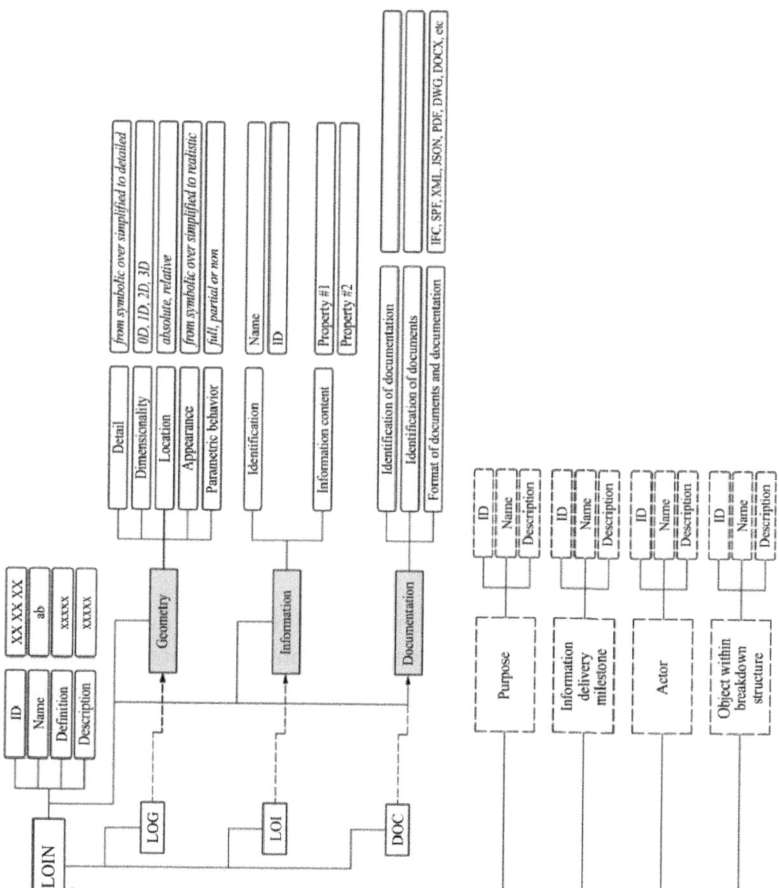

Figure 3.1. *Level of information needed framework (first version) from EN 17412-1 (2020) (LOIN: Level of Information Need; LOG: Level of Geometry; LOI: Level of Information; DOC: Documentation)*

The BIM environment makes it possible to move from a set of documents, whose interferences are difficult to manage, to a digital model (entity-association diagram). The model distinguishes between entities (business objects with properties and a manager) and their associations, and even the impacts on their immediate environment.

For this, the exchange should be specified to consider the lifecycle phase (Tolmer et al. 2015; Tolmer 2016). This is the notion of level of information need (LoIN) in ISO EN 17412 (2020) defining the level of information needed (geometric detail, associated documentation, actors involved, deadlines, etc.).

In an OpenBIM context where no single software covers all modeling and simulation needs, it is inevitable to use many specialized business software. This requires the ability to exchange information between different software, in order to avoid re-entries, sources of errors and digital discontinuity.

3.2.2. *Exchanges between modeling software*

A modeling software allows the describing of the 3D geometry of the different components of the structure, their geometrical and non-geometrical characteristics (materials, brand, technical characteristics, etc.), as well as their links or interfaces between components.

The native format of the software contains the model as described by the user, enriched with data specific to the optimization of the modeling and thus the know-how of its author. The information that links the objects together is based on geometry, often parametric geometry. This intelligence is intrinsic to the design software and therefore to the native format, for which there is a great risk of losing information when exchanging between different software, following the export in a neutral format that often only contains a limited collection of object instances.

3.2.3. *Exchanges between modeling and simulation software*

Analytical simulation software allows for the simulation of the behavior of a structure with respect to a given criterion. A calculation model is necessarily based on the geometry resulting from the modeling.

Simulation software includes different software – for example, for a roadway, planning software, acoustic or thermal analysis software, road traffic software or visibility control software.

As between modeling software via a neutral format, exchanges between modeling and simulation software are subject to the risk of information losses, due to simulation constraints leading to adaptations or simulations of the analytical model, compared to the accuracy of the geometric model.

3.2.4. *Exchanges between modeling software and other software*

These exchanges concern the information necessary for remote operation, topography, positioning of machines or construction tools, location of components, machine tool control, monitoring of construction and delivery of work. These are guidance instructions exported in specific formats, standardized or not standardized.

3.2.5. *Visualization software*

Visualization software allows for a geometric representation of the work under various points of view (contextual navigation) by means of navigation functions within the model. They give access to all the information of the selected object and sometimes perform certain operations automatically (dimensioning, annotations, recording of a point of view, etc.).

3.3. The sustainability of the information

Information is sustainable if it remains accessible over time and remains understandable.

3.3.1. *The security of standards*

The duration of the operation and maintenance of a structure requires information that is understandable and accessible for several decades.

Today, each publisher manages its own information format to optimize software performance. These proprietary formats evolve rapidly in order to support product improvements in the competitive world of IT.

This is contrary to the stability that is essential for users exchanging contractual data, which must be frozen to reflect a persistent state and which must be reused over long periods of time, specific to the construction and maintenance of the works.

The solution to this contradiction is the use of an international standard, which is the only way to ensure the durability of an information format and the control of its evolution.

The portable document format is an example of this. Developed by Adobe, the ISO-32000 (2017) standardization was imposed by the American Congress as a sine qua non condition for its use in order to digitize the documents of Congress and to ensure their conservation and their reliability of diffusion over a very long period.

3.3.2. *The storage of digital data*

In order to ensure the accessibility of information, several solutions (Benning and Cauvin 2018) exist:

– an identical backup or duplicate of the data to be able to restore it in case of damage or loss;

– an archive or copy of a data set, necessary for reference purposes.

However, ensuring accessibility is not enough. Ownership of the information and associated liability must also be managed.

A collaborative BIM database contains huge volumes of information generated by many contributors. Database laws should now apply to BIM as a composite work, owned by the author who made the integration, subject to the copyright of the pre-existing work (Benning et al. 2019a).

Intellectual property does not come under the technical provisions of the contract and must therefore be specified in a contractual document detailing the administrative clauses of the project. In any case, it remains clear that only the data concerning the finished work are impacted by these provisions. For example, the know-how and construction methods remain the responsibility of the engineering or the construction company, and as such, remain their property (Benning et al. 2019a).

It is therefore important to define each final deliverable and its content.

3.4. The development of IFC, a neutral exchange format

In order to meet all the needs expressed above, that is, an open object-oriented standard capable of facilitating data exchanges between software specific to the construction sector and ensuring their sustainability, the International Alliance for Interoperability (IAI), comprising several software publishers, was created in 1996 in order to develop the IFC standard, whose interoperability principle is described in Figure 3.2. In 2008, in order to communicate the objectives of the organization, IAI became buildingSMART International (bSI).

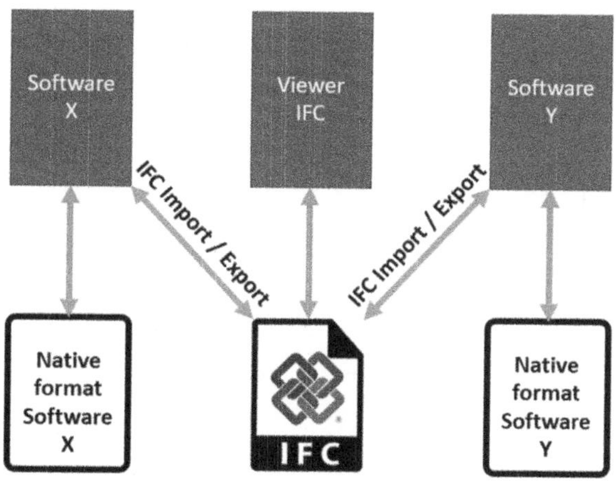

Figure 3.2. *IFC interoperability principle*

In ISO standardization, IFC (ISO 16739-1 2018) is based on the EXPRESS definition and its EXPRESS data definition language (ISO 10303-11 1994a; Schenck et al. 1994), the implementation of which uses the STEP physical file format (ISO 10303-21 1994b). The IFC Implementation Guide (Liebich 2009) provides examples of files written using the STEP physical file syntax.

3.4.1. *Principles, concepts and methods*

3.4.1.1. *Principles*

The structure of the IFC format is based on three main pillars:

– object classes;

– relationships between objects (such as facilities, spaces, areas, layouts, structural elements, connection to other objects);

– objects.

The characteristics of each object are themselves broken down into two categories:

– object class attributes, that is, metadata (shape, cost, position, energy performance, physical and mechanical properties, etc.);

– object properties, that is, the value of the attributes.

An example of an IFC object and possible relationships between objects are shown in Figure 3.3.

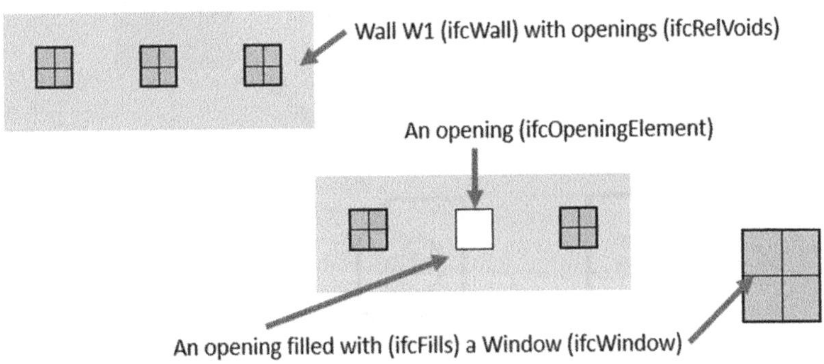

Figure 3.3. *IFC objects and relationships between objects*

The file is also composed of three parts:

– the header, containing general information about the file and the software used to generate it (example given in Figure 3.4);

– the data block, containing the description of a construction (components, geometries, technical characteristics, positions within the structure, etc.). Figure 3.5 is an example of a data block;

– the closing of the file.

Each line in the IFC file is structured in the same way: a unique number to designate the object, a reference to an object class and then, in parentheses, the values of the object's attributes. All these arguments are specified in the IFC documentation, as shown in Figure 3.5.

Building	Bridge	Civil works
IfcProject	IfcProject	IfcProject
IfcSite	IfcSite	IfcSite
IfcBuilding	IfcBridge	IfcFacility
IfcBuildingStorey		
IfcZone	IfcBridgePart	IfcFacilityPart
IfcSpace	IfcSpace	IfcSpace
IfcBuildingElement	IfcBridgeElement	IfcCivilElement

Figure 3.4. *Hierarchy of spatial division (building/infrastructure)*

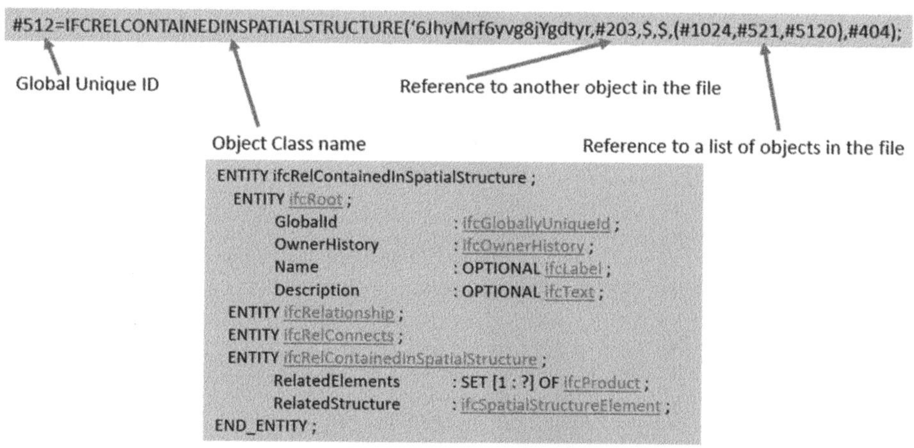

Figure 3.5. *The semantics of an IFC object*

3.4.1.2. *Entities*

The IFC defines an entity relationship model with several hundred entities organized in an object-based inheritance hierarchy (example given in Figure 3.6). Therefore, the entities include the following:

– building elements, such as *IfcWall*, defining a wall;

– geometries, such as *IfcExtrudedAreaSolid*, describing an extruded solid;

– basic constructions, such as *IfcCartesianPoint*, characterizing a geometrical point in a Cartesian frame.

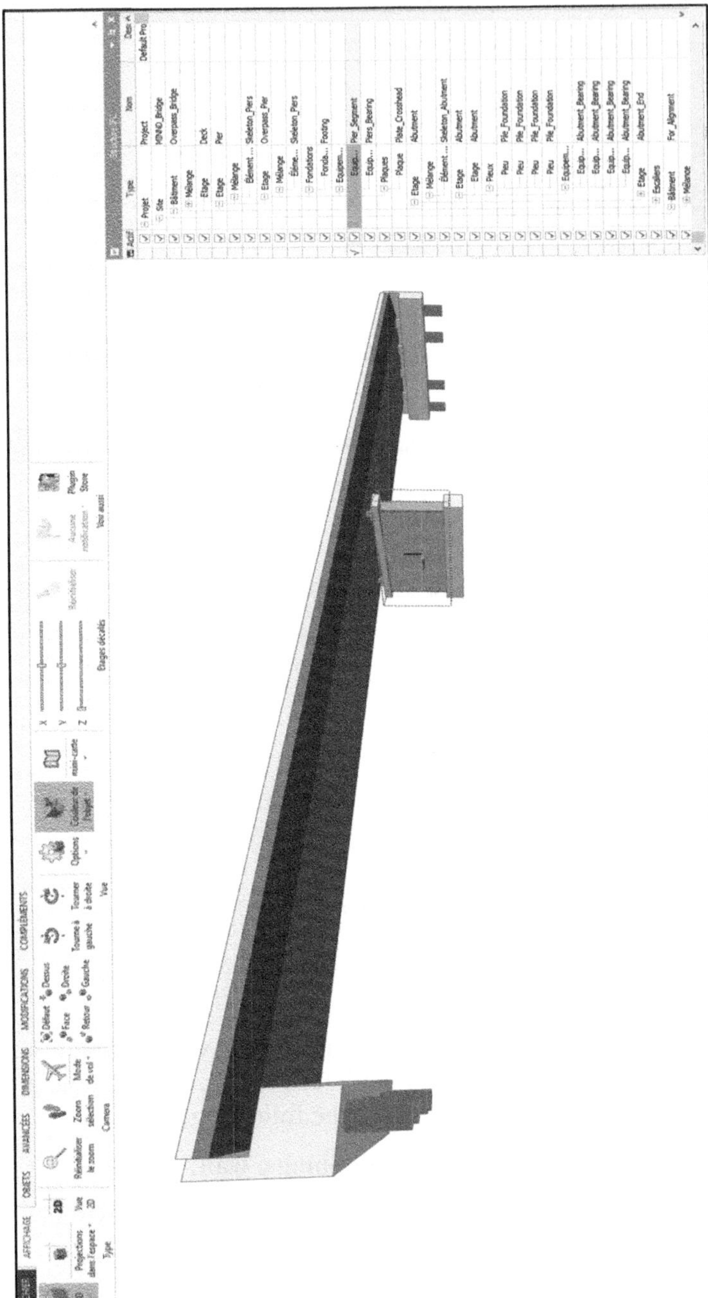

Figure 3.6. *The structure of IFC objects*

An entity is an object uniquely defined in the IFC model, characterized by a name (such as *IfcWall*) and attributes, themselves IFC entities. One of the methods of enriching the IFC model consists of adding attributes to an existing entity and renaming it to give it more detail: this is called derivation (see Figure 3.7).

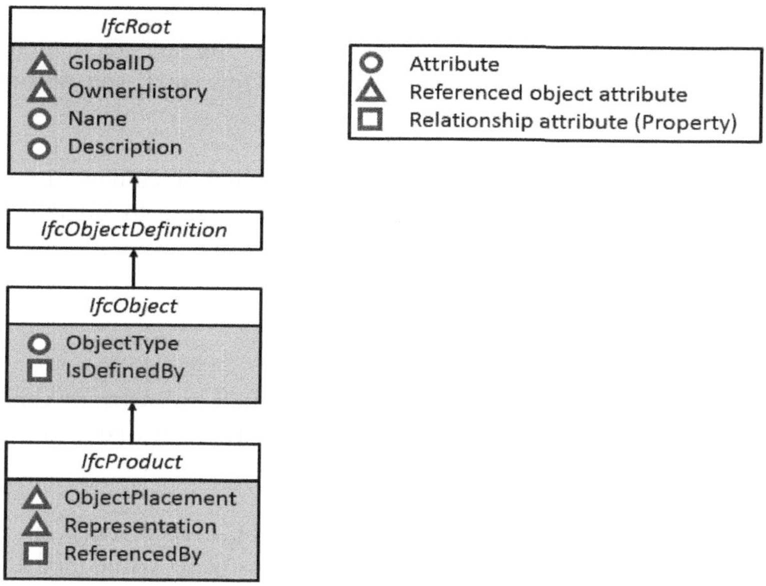

Figure 3.7. *Example of a derivation*

IfcRoot is a particular entity with the attributes of a unique identifier, its owner and the history of its modifications. All entities derived from *IfcRoot* have at the very least these same attributes, allowing them to be traced in exchanges.

All other entities are only potential attributes used to describe entities derived from *IfcRoot*.

IfcRoot is broken down into three abstract concepts:

– *IfcObjectDefinition* defining tangible objects;

– *IfcRelationship* describing the relations between objects;

– *IfcPropertyDefinition* detailing the properties of objects.

3.4.2. *Open format versus readability*

IFC is considered open format, that is, written in a non-encrypted language (and thus readable by a text editor), allowing it to be interpreted by any software, freeing it from the limits imposed by editors and their proprietary formats, the use of which is subject to licensing costs.

The reading and understanding of an IFC file are facilitated by a relatively explicit syntax (little compression, few text substitutions by alphanumeric codes, etc.).

Finally, bSI maintains a complete and detailed documentation on its technical site (buildingSMART 2020) that allows one to understand the syntax and to take part in training.

3.4.3. *IFC4*

The latest official version is IFC4, implemented by software vendors in 2020. A list of import/export compatible software is available on the bSI website.

Designed as a continuation of IFC2x3, IFC4 represents an overall improvement in model quality. Most of the classes were maintained and only 113 were added (766 total vs. 653 for IFC2x3) (Benning et al. 2018).

Nevertheless, this new version offers some notable advantages:

– a better concordance of interpretations (all similar concepts are modeled in the same way);

– a separation between general and parametric definitions;

– an exhaustive list of object-types;

– the possibility of energy analysis and carbon balance;

– integration of environmental data;

– greater flexibility in modeling shapes: extension of geometric representations to model more complex shapes (non-uniform rational B-spline representation);

– simpler visualization of objects with tessellated geometry introduced to simplify the representation of complex shapes.

3.4.4. *Other related formats*

3.4.4.1. *CityGML*

CityGML (OGC 2012) is an open data model and XML-based format addressing the storage and exchange of virtual 3D city models. It is an implementation scheme of GML3 (Geography Markup Language, Version 3.1.1), an extensible international standard for geographic data exchange established by the Open Geospatial Consortium (OGC) and ISO/TC211 (1994). The objective of the development of CityGML is to achieve a common definition of the basic features, attributes and relationships of a 3D city model. This is particularly important for the cost-effective and sustainable maintenance of 3D city models, allowing the reuse of the same data in different fields of application.

3.4.4.2. *LandXML*

LandXML (2018) specifies an XML file format for civil engineering design and topography measurement data. The main objectives for providing a standard data format are as follows:

– data exchange between software;

– long-term data archiving.

Hundreds of software developers and government organizations around the world have adopted LandXML.

3.4.4.3. *InfraGML*

InfraGML (OGC 2016) is the OGC application schema for land development and civil engineering infrastructure. It supports a subset of the existing LandXML standard.

3.5. The infrastructure domain

3.5.1. *Definitions*

Here, “infrastructure” is employed in the sense of land use, urban planning and transportation infrastructure. An infrastructure is a “set of facilities built on the ground or underground that allow for human activities to be carried out across space” (Merlin and Choay 2015, author’s translation). It is composed of earthworks, bridges and tunnels, supporting a superstructure (roadway, railroad, etc.). It is often linear and associated with a right-of-way and dependencies (connections to existing structures).

An infrastructure strongly impacts its environment, modifying an existing space by removing and then adding material. The current practice only describes a finished structure, whereas BIM and IFC make it possible to define the evolution of the structure during its lifecycle (delivered structure and temporary structures necessary for the works in question).

3.5.2. *Specificity of the infrastructures*

Designing and building an infrastructure requires taking into account the following specificities:

– scales of units varying from millimeters to kilometers, or even to a hundred kilometers;

– an exact knowledge of the context (buildings, networks, biodiversity, water, etc.);

– knowledge of the subsoil (natural terrain, geology, hydrology, geotechnics);

– georeferencing of the objects (in a coordinate system and geodetic projections) (see Figure 3.8) and the use of a linear referencing system to position the structures along the curvilinear abscissa of the reference axis;

– intermediate phases of construction (construction methods often dimension the structure, temporary structures, coactivity of cumbersome trades, etc.);

– complex analytical models linked to geometrical models, where the geometry of some objects is defined by calculation (e.g. suspension and stay cables of a bridge, stays of a structure).

3.5.3. *BIM challenges for infrastructure*

In order to take into account the specificities of infrastructures, the design and simulation models must face the following challenges:

– the uncertainty of geotechnics, with the notions of associated risks;

– the need for complementarity with the Geographic Information System, which allows numerous data concerning the environment of the territory and the context of the works to be realized;

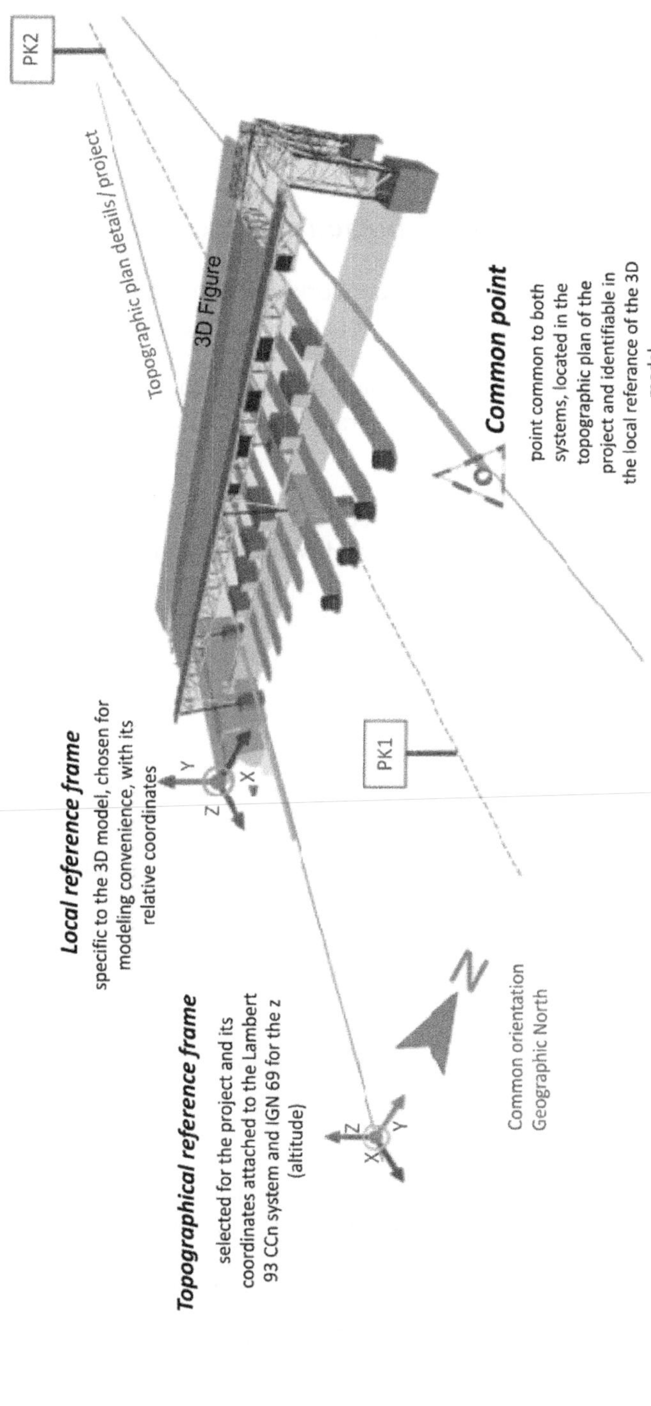

Figure 3.8. *The georeferencing of objects (Benissan 2019)*

– the interfaces between systems and disciplines in order to guarantee the continuity of service and the restoration of networks;

– the consideration of post-construction phases, that is, operation and maintenance, whose major maintenance strategies are linked to construction methods.

3.5.4. *Comparison with the manufacturing industry*

An infrastructure is always a unique work, built in place, whose design is rarely completed and validated before construction begins.

Moreover, each project is carried out in a specific contractual environment, most often in a public procurement framework, where the partnership is systematically adapted to the project.

The similarities with the manufacturing industry concern the rise of industrialization to limit heavy labor and increase productivity and quality (use of robots), but also the increasingly frequent use of heavy prefabricated elements to reduce deadlines and to free oneself from hazards linked to the weather or to the constrained physical environment.

3.6. IFCs for infrastructure

3.6.1. *Identified areas*

The bSI InfraRoom has initiated several working groups on defining IFCs for infrastructure (Sacks et al. 2018).

The buildingSMART International IFC roadmap for infrastructure (Figure 3.9) is broken down into several areas.

The first domain identified was IFC-Alignment, which allows for the localization of linear project objects in a reference linked to the curvilinear abscissa of the reference axis.

The next domain concerns engineering structures (*IFC-Bridge*: Benning 2019; buildingSMART 2020) and initially common structures whose constitution is quite similar to that of buildings. For more complex structures (with the use of prestressing components or load-bearing cables), the definition of the particular components will be specified later.

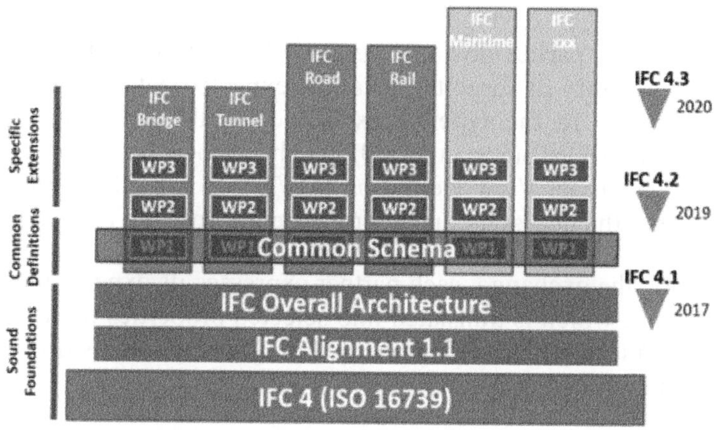

Figure 3.9. *buildingSMART International Roadmap Diagram, an IFC roadmap for infrastructure*[1]

The following domains are under development:

– *IFC-Tunnel*, including structure (civil engineering) and geotechnics/ geology;

– *IFC-Road*, including objects associated with roadways and earthworks;

– *IFC-Rail*, including all the objects required for railroads (track, signaling, telecommunications, etc.).

Other areas will be addressed when international experts are mobilized and organized.

3.6.2. *Development methodology*

The exchange of information is based on entities specific to the vocabulary of the infrastructure business.

Establishing a dictionary of the business data to be exchanged is the first step toward interoperability. Although it must be multilingual, English is the working language of the data dictionary (DD). It defines the entities of the IFC data model and their main properties. It must be compliant with the ISO 23386 (2020) product description standard.

1 https://www.buildingsmart.org/standards/calls-for-participation/ifcroad/ [Accessed 15/02/2023].

The content of the exchange depends on the phase of the lifecycle and the expectations of the parties involved, as not all data in the DD are involved. To avoid unnecessary or forgotten data, the expected data are specified in a document specific to the type of exchange, in the English Information Delivery Manual (IDM) (See et al. 2012). The associated subset of the IFC model is called Model View Definition (MVD). It is used to filter the data and deliver only the necessary and appropriate information to each partner.

To describe the exchange, each partner expresses its expectations with its own vocabulary, which is translated into selected properties in the structured data set. The filtering of information according to a partner's discipline is a matter of systems engineering.

3.6.3. *Newly built classes*

3.6.3.1. *IFC-Alignment*

An infrastructure carries lanes where vehicles travel. The environment and the traffic conditions impose a specific alignment of the lanes, called reference axis or alignment. IFC-Alignment allows for describing these alignments in the classical orthonormal reference frame as well as the local reference frames associated with the alignments and the vehicles.

The IFC-Alignment-4.1 conceptual model is now available and being implemented by software editors (buildingSMART 2020).

3.6.3.2. *IFC-Bridge*

A bridge carries traffic lanes whose axes are managed by alignments. New classes have been added, allowing the management of extruded geometries along alignments and describing the geometry of decks and piers. Prestressing cables, dense reinforcement of concrete elements and specific supports have been integrated.

The interaction with the ground (deep foundations, access embankment) will be treated with tunnels (IFC-Tunnel) and earthworks (IFC-Road).

The IFC-Bridge-4x2 conceptual model is available and being implemented by software vendors (Benning 2019; buildingSMART 2020).

3.6.4. *Classes under development*

3.6.4.1. *IFC-Tunnel and IFC-Geotechnics*

A tunnel houses traffic lanes. The description of the geometry of the structure is based on the application of alignments to bridges. The modeling of the numerous networks necessary for the operation and safety of the tunnel is based on the modeling of the buildings. The strong impact of geology and geotechnics requires specific developments that can be used for other structures (bridge, building, etc.).

3.6.4.2. *IFC-Road and IFC earthworks*

Roads benefit from bridge and tunnel work. The modeling of earthworks requires new developments that will benefit other infrastructures.

3.6.4.3. *IFC-Rail*

The rail system is based on the work, done or in progress, concerning the infrastructure. The new developments concern the modeling of the track and the equipment necessary for operation (energy, signaling, telecommunications).

3.6.5. *Perspectives*

The major challenges ahead are as follows:

– The integration of all the infrastructure domains, in order to verify the completeness and the non-redundancy of all the work carried out by the different international groups of experts, but also the objects common to all the domains (drainage, earthworks, etc.).

– The link between the architectural model (geometric) and the calculation models (analytical), which are necessary in certain cases (see Figure 3.10). For example, the load-bearing cables of bridges, whose geometry is defined by the forces applied to the cables. The general conceptual model of IFCs must then be enriched to take into account these new requirements (Benning et al. 2018).

– The inclusion by IFC for the concept of a functional diagram necessary for the understanding and operation of the works (Vanlande 2008). Currently, the backlogs of executed works from the IFC models are a collection of geometric models and documents describing the equipment implemented. However, the operation of a structure is essentially based on the knowledge and operation of systems described by functional diagrams.

– The management of modifications and the traceability of project evolutions, which are already implemented in IFCs, but are little used, even though this is one of the intrinsic assets of the IFC conceptual model.

The approach described by the IFC is based on a description of the objects; however, one needs to understand that within the cooperative modeling approach of BIM, if it is necessary to know how to describe a world of objects, then it is not a static vision that is needed, but *a vision of the processes*:

> We can think of the world as made up of things. Substances. Entities. Of something that is. That remains. Or we can think of the world as made up of events. Processes. Of something that happens. That does not last, that is continuously transformed. Thinking of the world as a set of processes is the mode that allows us to better understand and describe it. Thinking of the world as a set of processes is the mode that allows us to better understand and describe it. The world is not a set of things, it is a set of events. (*The Order of Time*, Carlo Rovelli, chapter 6, author's translation from the French edition)

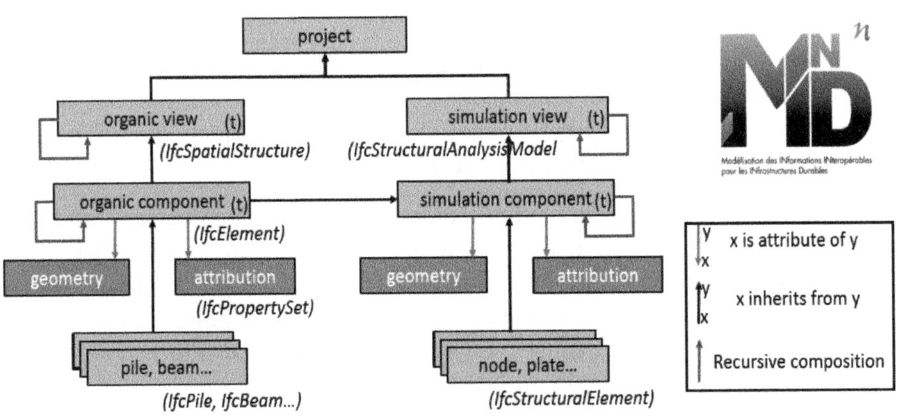

Figure 3.10. *The data model for the architecture/simulation link. For a color version of this figure, see www.iste.co.uk/teulier/building.zip*

3.7. Standards

3.7.1. *IFC standards*

The main IFC standards are listed below:

– ISO 16739-1 (2018) – Industry Foundation Classes (IFC) for data sharing in the construction and facility management industries – Part 1: Data schema;

– ISO 10303 (STEP) AP 239 edition 3 – Application Protocol for Product Life Cycle Support (PLCS);

– ISO 29481-1 (2010) – Building Information Modeling – Information Delivery Manual – Part 1: Methodology and Format (IDM) (in order to have a methodology to capture and specify processes and information flow during the lifecycle of a facility);

– ISO 12006-2 – Building Construction – Organization of Information about Construction Works – Part 2: Framework for Classification;

– ISO 12006-3 – IFD – Organization of Information about Construction Works – Part 3: Framework for Object-Oriented Information. Data Dictionaries.

3.7.2. *BIM and related standards*

The main BIM and related standards are listed below:

– ISO 19650-2 (2018) Organization and digitization of information about buildings and civil engineering works, including building information modeling (BIM) – Information management using building information modeling – Part 2: Delivery phase of the assets;

– ISO 19650-3 (2020) Organization and digitization of information about buildings and civil engineering works, including building information modeling (BIM) – Information management using building information modeling – Part 3: Operational phase of the assets;

– ISO 19650-1 (2018) Organization and digitization of information about buildings and civil engineering works, including building information modeling (BIM) – Information management using building information modeling – Part 1: Concepts and principles;

– ISO 19650-2 (2018) Organization and digitization of information about buildings and civil engineering works, including building information modeling (BIM) – Information management using building information modeling – Part 2: Delivery phase of the assets;

– ISO 19650-3 (2020) Organization and digitization of information about buildings and civil engineering works, including building information modeling (BIM) – Information management using building information modeling – Part 3: Operational phase of the assets;

– ISO/CD 19650-4 (2020) Organization and digitization of information about buildings and civil engineering works, including building information modeling (BIM) – Information management using building information modeling – Part 4: Information exchange;

– ISO 19650-5 (2020) Organization and digitization of information about buildings and civil engineering works, including building information modeling (BIM) – Information management using building information modeling – Part 5: Security-minded approach to information management;

– ISO 23386 (2020) Building information modeling and other digital processes used in construction – Methodology to describe, author and maintain properties in interconnected data dictionaries;

– EN 17412-1 (2020) Building Information Modeling – Level of Information Need – Part 1: Concepts and Principles;

– EN ISO 16739-1 (2018) Industry Foundation Classes (IFC) for data sharing in the construction and facility management industries – Part 1: Data schema.

3.8. References

Benissan, C., Benning, P., Cousin, V., Guilloteau, S., Luzik, Y., Page, J.F., Rives, M. (2019). Recommandations de mise en place du BIM pour les infrastructures [Online]. Available at: https://www.minnd.fr/wp-content/uploads/2020/09/MINnD_TH01_UC00_01_Recommandations_BIM_Infra_024_2019.pdf [Accessed 6 May 2021].

Benning, P. and Cauvin, B. (2018). Bridge data dictionary from conception to bsDD implementation [Online]. Available at: https://www.minnd.fr/wp-content/uploads/2020/09/MINnD_UC03_02_Bridge_data_dictionary_from_conception_to_bSDD_015A_2018.pdf [Accessed 30 April 2021].

Benning, P., Cauvin, B., Cousin, V., Dumoulin, C., Fliscounakis, A., Ledan, A. (2018). Bridge design process & IFC extensions MINnD [Online]. Available at: https://syncandshare.lrz.de/download/MktrQlVFTm1CTHBLS3lNcW5MQVdz/IFC%20Bridge/20170123_Auftakttreffen/MINnD_Tr2-BridgeDesignProcess_v2.2.2.pdf?inline [Accessed 13 December 2020].

Benning, P., Ayraud, S., Cousin, V., Demilecamps, L. (2019a). La question de la propriété intellectuelle. In *Le BIM entre recherche et industrialisation*, Tolmer, C.-E. and Teulier, R. (eds). Eyrolles, Paris.

Benning, P., Cauvin, B., Corre, G., Cousin, V., Dumoulin, C., Ledan, A. (2019b). IDM bridge design process and IFC extensions MINnD [Online]. Available at: https://www.minnd.fr/wp-content/uploads/2020/09/MINnD_UC03_03_IDM_Bridge_design_process_015B_2019.pdf [Accessed 18 November 2020].

buildingSMART (2020). Industry foundation classes version 4.2 bSI draft standard IFC bridge proposed extension [Online]. Available at: https://standards.buildingSMART.org/IFC/DEV/IFC4_2/FINAL/HTML/ [Accessed 15 January 2021].

Dumoulin, C. (2018). Indicateurs de l'usage du BIM – Livrables Saison 1 – Observatoire [Online]. Available at: www.minnd.fr [Accessed 15 January 2021].

EN 17412-1 (2020). Building information modelling – Level of information need – Part 1: Concepts and principles [Online]. Available at: https://www.en-standard.eu/bs-en-17412-1-2020-building-information-modelling-level-of-information-need-concepts-and-principles/ [Accessed 15 February 2023].

LandXML (2018). New version of free LandXML viewer adds point clouds [Online]. Available at: http://www.landxml.org/ [Accessed 14 January 2021].

Liebich, T. (2009). *IFC Model Implementation Guide*, 3rd edition. buildingSMART Model Support Group.

Merlin, P. and Choay, F. (2015). *Dictionnaire de l'urbanisme et de l'aménagement.* PUF, Paris.

Norme ISO 10303-11 (1994a). Industrial automation systems and integration – Product data representation and exchange – Part 11: Description methods: The EXPRESS language reference manual. International Organization for Standardization, Geneva.

Norme ISO 10303-21 (1994b). Industrial automation systems and integration – Product data representation and exchange – Part 21: Implementation methods: Clear text encoding of the exchange structure. International Organization for Standardization, Geneva.

Norme ISO 10303-239 (2012). Industrial automation systems and integration – Product data representation and exchange – Part 239: Application protocol: Product life cycle support. International Organization for Standardization, Geneva.

Norme ISO 32000-2 (2017). Document management – Portable document format – Part 2: PDF. International Organization for Standardization, Geneva.

Norme ISO 16739-1 (2018). Industry Foundation Classes (IFC) for data sharing in the construction and facility management industries – Part 1: Data schema. International Organization for Standardization, Geneva.

Norme ISO 19650 (2018). Organization and digitization of information about buildings and civil engineering works, including building information modelling (BIM) – Information management using building information modelling. International Organization for Standardization, Geneva.

Norme ISO 23386 (2020). Building information modelling and other digital processes used in construction – Methodology to describe, author and maintain properties in interconnected data dictionaries. International Organization for Standardization, Geneva.

Norme ISO/TC211 (1994). Geographic information/Geomatics. International Organization for Standardization, Geneva.

OGC (2012). CityGML [Online]. Available at: https://www.ogc.org/standards/citygml/ [Accessed 10 December 2020].

OGC (2016). Land and infrastructure (LandInfra) [Online]. Available at: https://www.ogc.org/standards/landinfra/ [Accessed 5 December 2020].

Rovelli, C. (2018). *L'ordre du temps*. Flammarion, Paris.

Sacks, R., Eastman, C., Lee, G., Teicholz, P. (2018). *BIM Handbook: A Guide to Building Information Modeling for Owners, Designers, Engineers, Contractors, and Facility Managers*, 3rd edition. John Wiley & Sons, Hoboken.

Schenck, D. and Wilson, P. (1994). *Information Modeling the EXRESS Way*. Oxford University Press, New York.

See, R., Karlshoej, J., Davis, D. (2012). An integrated process for delivering IFC based data exchange, International User Group, buildingSMART International [Online]. Available at: https://standards.buildingSMART.org/documents/IDM/IDM_guide-IntegratedProcess-2012_09.pdf [Accessed 5 November 2020].

Tarandi, V. (2011). The BIM collaboration HUB: A model server based on IFC and PLCS for virtual enterprise collaboration. In *Proceedings of the CIB W78-W102 2011: International Conference*, Sophia Antipolis.

Tolmer, C.E. (2016). Contribution à la définition d'un modèle d'ingénierie concourante pour la mise en œuvre des projets d'infrastructures linéaires urbaines : prise en compte des interactions entre enjeux, acteurs, échelles et objets. PhD Thesis, Université Paris Est, Paris.

Tolmer, C.E., Castaing, C., Morand, D., Diab, Y. (2015). Modèle conceptuel pour la structuration de l'information en conception d'infrastructures. In *Rencontres Universitaires de Génie Civil*, May, Bayonne.

Vanlande, R., Nicolle, C., Cruz, C. (2008). IFC and buildings lifecycle management. *Automation in Construction*, 18, 70–78.

4

Structuring Information for the Digital Twin

Ana Roxin[1], Christophe Castaing[2,3,4] and Charles-Édouard Tolmer[5,4]

[1]Laboratoire d'Informatique de Bourgogne, Université Bourgogne-Franche-Comté, Dijon, France
[2]buildingSMART International, Paris, France
[3]Groupe Egis, Paris, France
[4]Projet MINnD, Paris, France
[5]Eurovia Infra, Vinci Construction, Nanterre, France

4.1. Introduction

As explained in the collective work by Bot and Vitali (2011), the design world is currently undergoing a paradigm shift. This shift is because of advancements made in techniques and science rather than a leap within a continuity. Currently, building information modeling (BIM) is the only consequence of this paradigm shift, facilitated partly by the rapid arrival of new technologies and more powerful tools (design tools but also data, graph-oriented databases, ontologies, etc.). This context imposes but allows for the mobilization of holistic knowledge when the designer joins, brings together, combines, arranges and "synthesizes" their design activity (Bot and Vitali 2011). In their work, Micaelli et al. (2011) discuss three paradigms of the design domain. In contrast to the design paradigms that are "artisanal and empirical", the key elements of the "abstract design" paradigm are as follows:

– the notion of model and modeler is still central: designing is modeling;

– the coproduction, by multidisciplinary teams, and the circulation of models between teams, whether present or remote, are emphasized;

– the distribution of work is done through codified processes and not in a Taylorian way as in empirical design;

– the use of generic and prescriptive abstract models from which concrete processes and artifacts are generated becomes central.

In this abstract design paradigm, design is based on a particular tool, which is neither a testing ground, a mock-up, nor a schematic, but instead a generic model, capable of describing both the artifact (referred to as the system "to be made" in this chapter's context) and its design process (referred to as the system "to make") (Bot and Vitali 2011). Azhar (2011) describes BIM as the new paradigm for the construction industry. This reflects a widely held view in the construction profession. BIM as a methodology is also seen as bringing together, for projects, the consequences of the paradigm shift (Tolmer 2016). Among these consequences, we may cite new needs in modeling and formalisms, the project's vision as a system, and requirements management. They require global thinking of project information, notably through data models, improvements in interoperability, and so on.

4.2. Problem

As digital processes are implemented and generalized in the act of building or maintaining, a vision of structuring data versus information is required because the digital twin or virtual asset is imposed on the owner as one of the deliverables needed to meet the complete management of the lifecycle of a work. The asset is to be understood in the patrimonial sense, as stated in ISO 19650-1[1]: "item, thing or entity that has potential or actual value to an organization" (2018, 3.2.8). This virtual asset imposes the need for a "scientific" approach to address transforming data into knowledge and implementing the processes that enable interoperability between the underlying models of that knowledge. How can we obtain relevant

1 To get definitions of ISO standards free of charge, see the Online Browsing Platform (OBP) at https://iso.org/obp/ui..

information from all the data available for a structure? How can it be modeled, and how can knowledge be extracted from it? These are some of the questions that this chapter attempts to answer. To obtain efficient digital processes that are as faithful as possible to the realities on the ground, it is necessary to implement interoperability between information models. This chapter examines the links between models and their standardization toward interoperability, all applied to the context of the digital twin.

The digital twin (or virtual asset) is an information system that transforms unstructured data into structured information around which services are developed to enable exploitation. The first service is sharing, in the four states of information, as defined in the first part of the ISO 19650-1 (2018) standard. A civil engineering structure must meet a primary requirement: durability, both in the sense of "durability" and "sustainability". However, data and information (see definitions below) do not naturally possess this property. To obtain a digital description of a built structure, it is necessary to be able to match data coming from several sources: different technical domains, different design tools, different geometrical references and different structuring approaches. This matching need is referred to as the need for interoperability. ISO 17261 defines interoperability as the "ability of systems to provide services to and accept services from other systems and to use the services so exchanged to enable them to operate effectively together" (2012, 3.24). From a machine point of view, implementing interoperability means linking two heterogeneous computer systems to collaborate, which implies reciprocal access to their resources. Interoperability has no reason to meet the criterion of durability: diversity, on the one hand, and tools' lifecycle, on the other, do not naturally fit into this requirement. Standardization is, therefore, necessary to ensure data interoperability and the system of relationships implemented to obtain an information system. The purpose of this chapter is to show which tools are available to structure information around models (geometric and semantic) that allow not only to describe the works to be built or already built fully, but mostly to develop one or more necessary models, that, once standardized, will be open and neutral. In this sense, standardization work in this field goes far beyond simply creating a consensus on prior knowledge.

Two definitions are essential to note here:

– data are "output resulting directly from the measurement of variables" (ISO 772 2011, 1.133);

– information is a “reinterpretable representation of data in a formalized manner suitable for communication, interpretation or processing” (ISO 19650 2018, 3.31).

This coupling between data and information is part of what Ackoff (1989) calls the “knowledge pyramid” (Figure 4.1), which allows linking decision-making to a continuous schema of data > information > knowledge > actions (Figure 4.2).

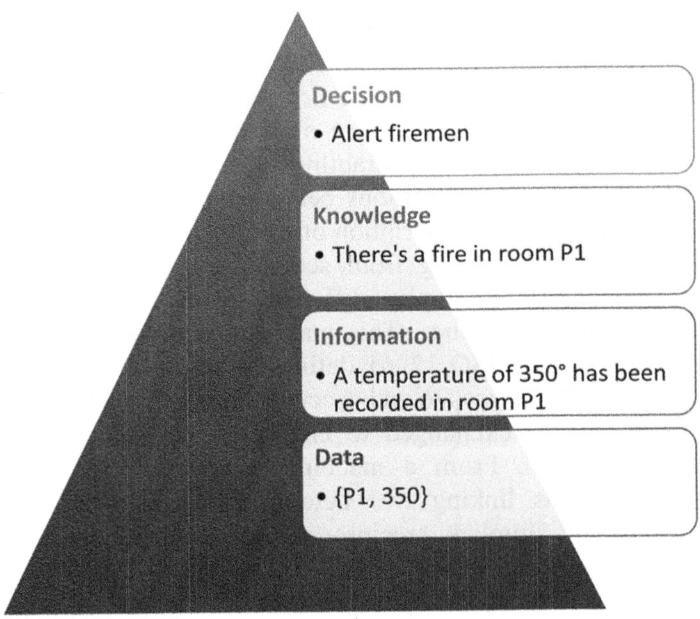

Figure 4.1. *Knowledge pyramid, adapted from Ackoff (1989)*

The digital twin must contain both static and analytical models (also called dynamic models), which are two different kinds of abstractions (an abstraction is modeling and therefore necessarily a simplification) of the same reality, that of the structure and its environment. Each model has its utility and degree of simplification concerning reality. However, the digital twin must ensure data, information and knowledge continuity and coherence in each model.

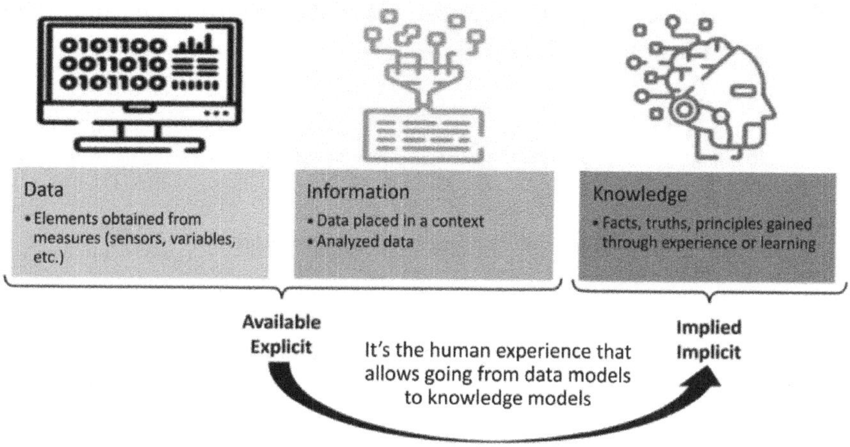

Figure 4.2. *From "data" to "knowledge" – this figure summarizes the heuristic that led the construction industry to reflect upon data structuring*

It turns out that, progressively and for several years, in the modeling work carried out by the construction industry, a junction is taking place with the manufacturing industry around systems engineering for two reasons:

– the need to decompose a construction work not only as objects but also as systems (not only manufacturing but also as conceptual objects) and spaces;

– the need to understand the building process, as a system in itself, for the production of the digital asset.

The digital twin relies on interoperability, which in turn relies on standardization. Indeed, three approaches exist to implement interoperability: models can be unified, integrated or federated (ISO 11354-1 2011). The digital twin first requires modeling (which allows for the structuring of data), next a normalization of models (which allows for obtaining information) and finally interoperability between models (which brings knowledge). For the latter, three possibilities exist (ISO 14258 1998): two models can be integrated, unified or federated.

This process, as well as the use of data, information and finally, knowledge, is illustrated in Figure 4.3.

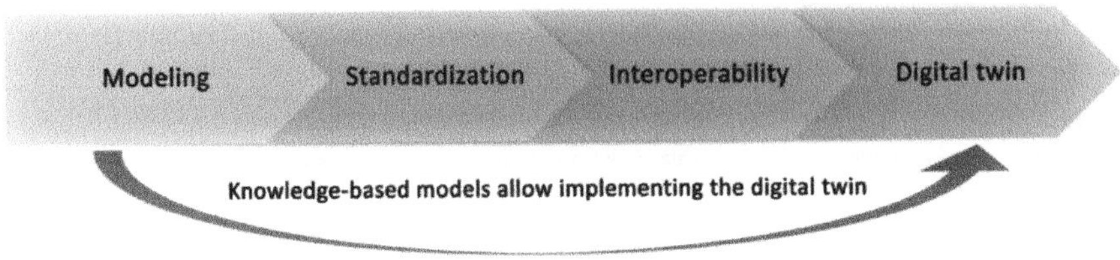

Figure 4.3. *From modeling to the digital twin*

Knowledge is used twice in this process:

– at the origin of the process to imagine the decomposition into objects and systems;

– at the end of the process, as a result, to give an account of the actual work carried out.

Describing the structuring of the data necessary to deliver a digital asset requires the heuristic contained in the following definition.

The digital twin is an information system based on the following elements:

– an organized progression between data and information;

– a description and organization of data within a context to structure them in the form of information;

– an ontology, which is its semantic modeling, that is, an explicit and formal specification of the related knowledge and its possible interpretation in a given context;

– a distinction and distribution between the conceptual model and the semantic model (ontology);

– semantic interoperability between models allows a computer to coherently interpret another system's resources and exchange knowledge with another computer system.

4.2.1. *Complex systems*

Current scientific results are significant for the modeling and control of technical systems. Nevertheless, efficient management of these systems is often beyond our reach and requires the design of specific software solutions. Current problems related to sustainable development and environmental preservation are complex and composite. Their study and modeling are based on multidisciplinary approaches and result in the definition of complex systems. Since complex systems are interwoven with many social, technological and natural processes and are connected to diverse organizations, there are no single solutions or approaches to modeling or designing them (MINnD 2019). We further clarify what we mean by "system", define a system's complexity and discuss the differences between complex systems and "systems of systems".

4.2.1.1. *Definitions and features of a system*

According to Levin, a "system" is "a set of interdependent or temporally interacting parts where parts are, generally, systems themselves, and are composed of other parts in turn" (Levin 2006). Annex D of ISO 15288 considers these systems to be "man-made, created and used to provide services in a defined environment for the benefit of users and other parties involved" (ISO 15288 2015).

Bunge's general systems theory (Bunge 1979) specifies the three components of a system, namely:

– the set of its parts represents the Composition (C) of a system;

– the Environment (E) of a system is the ensemble of entities linked to the parts involved in the system's Composition (C);

– the Structure (S) of a system contains the ensemble of internal (between the elements of the system's Composition C) and external relations (between elements of the system's Composition and the elements of the system's Environment E).

These three components bring the following three properties:

– the interconnectedness represents the set of external links of a system;

– the interconnectedness between the system and its environment creates vulnerabilities and risks that must be analyzed and exposed for system managers, sponsors and public policymakers;

– the interdependence is defined as the set of internal links in the system;

– the complexity is the number of parts of the system that interact with each other. The more interfaces there are between these parts, the greater the complexity of the associated system.

We retain two aspects: on the one hand, the notion of complexity is defined by the number of interfaces, and on the other hand, the notion of a system which aims at what is man-made.

4.2.1.2. *Complex systems and "systems of systems"*

A large number of interacting components characterize complex systems. Complex systems span multiple dimensions, including economic, ecological and social subsystems, which may have any level of interaction with each other. Requirements throughout the lifecycle address this complexity.

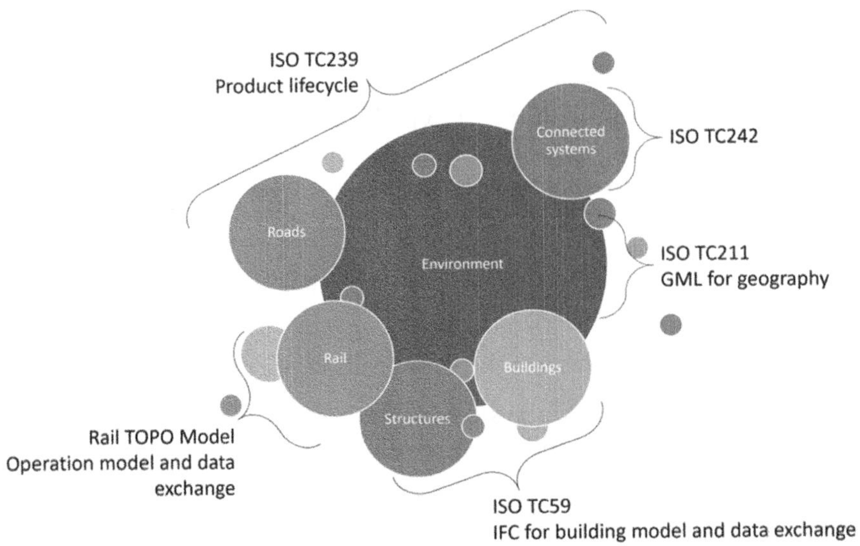

Figure 4.4. *Illustration for a particular case of complex system, for example, a system of systems. For a color version of this figure, see www.iste.co.uk/teulier/building.zip*

To better illustrate the concept of a "complex system", let us take the example of a railway project (Figure 4.4). In this context, complexity appears as a matter of course through the number of interfaces linked to the subsystems that compose it, for example, interfaces between the complex system "environment" and the complex systems "road" and "rail". Figure 4.4 also allows us to define systems of systems as a particular case of complex systems. Indeed, all the systems illustrated derive from the same organization, which structures the functioning of the railway system intrinsically. However, the environmental dimension is not derived from structuring a railway project. The "environmental system" has its own functioning and behavior rules and requires specific modeling. The usefulness of this decomposition forces us to think, within the framework of model-driven engineering, of models of a different nature, which the use of an ontology cannot regulate.

This approach also allows us to consider solutions to the problems of smart networks or smart cities since we need to consider not only the interaction of complex systems and subsystems but also the interaction with an environment that does not have a functional dependency on the systems of networks, or city, in question.

This remark allows us to understand that the treatment of these interactions requires going beyond the identification of the interface and treating the modeling of this environment with different conceptual models. It should be understood that the modeling of geology or fauna should be based on different rules, relations and representations than those used to represent an artifact.

In what follows, we present the business issues of modeling and discuss the main challenges associated with these. Finally, we present the main existing approaches at two levels: at a conceptual level and an implementation level. The following adheres to the second part of the heuristic: the vision of BIM as a starting point for creating the digital twin that forces us to think of the twin in its dynamic. Imagine the object to be built in its evolution throughout its lifecycle: understand the interaction and impacts of the design, construction and operation processes on the digital twin. We cannot only model the data but must also represent and model the processes.

4.2.2. The business issue: "enabling system" and "systems of interest"

We will see that systems engineering is not the only solution to the new design paradigm. Some authors (Rochet 2007; Fiorèse and Meinadier 2012) describe it more as a reflection tool, an open working method, and not a generic model to be derived for each project. It is also a "scientific evolution necessary to design sustainable urban ecosystems" (Rochet and Peignot 2013, author's translation).

Today, these complex adaptive systems are evolving to cope with uncertainties and changes in the operating environment, social values and technological advances. Growing human populations and increasing demand for limited natural resources require sustainable cities to use efficient energy production, water management and transportation infrastructure. Artificial intelligence (AI), machine learning algorithms, ubiquitous sensing and activation capabilities, real-time control, cloud computing, data science and analysis are among the main tools for these systems.

Given that, depending on the business domain under consideration, systems engineering is practiced in different ways, we present the issues underlying systems modeling in the context of infrastructure projects.

The digital twin must be understood as a representation, not only static but also dynamic, of a structure: from its origin, the digital representation must integrate the markers for monitoring its progress in the lifecycle. The information must be managed for adapted use and must, therefore, carry the data by qualifying it (the metadata). The organization in tree structures is inefficient, even fragile, compared to the metadata modeling that allows for access and more powerful traceability.

The development of BIM and the digital twin pushes the processes toward the "abstract design" mentioned above. The digital transition of the construction sector in project management introduces the valorization of information, structuring metadata in graphs and no longer in file trees, management of data dictionaries and product catalogs, etc. Others come from introducing new methods from other industries, such as requirements management, systems engineering, knowledge management and product lifecycle management. Modeling could be the common denominator to reconcile these paradigms and different conceptualizations and structuring of information. The use of standard, normalized or, failing that, standardized formalisms for modeling seems to be a promising prospect, according to numerous works and studies conducted in other industries.

The ISO 19650 (2018) digital twin standard (the Asset Information Model or AIM) is an example. It meets two objectives: representing the asset and bringing it to life. ISO 19650 (2018) describes two types of requirements: those relating to the "performance" of the asset (requirements of the asset, the product) and those on the performance of the project processes to contribute to the permanent and dynamic realization of the digital twin (Figure 4.5).

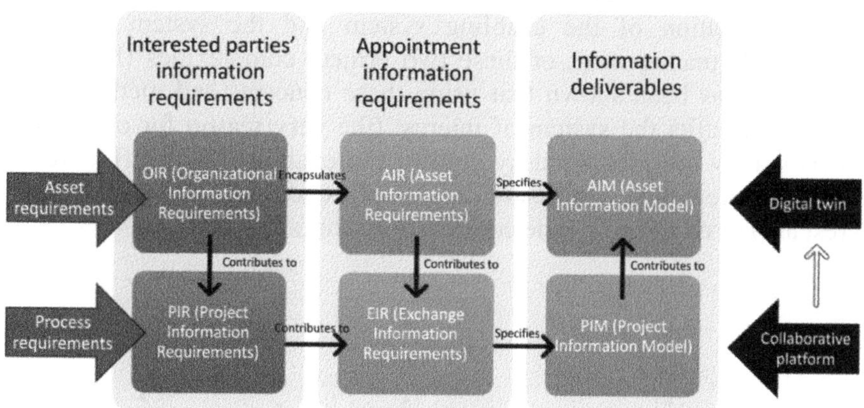

Figure 4.5. *Structuring provided by ISO 19650 (2018)*

ISO 19650 (2018) structures the object to be realized (the product or system of interest) and the processes, ensuring the lifecycle of the data and its qualification (the project or enabling system). The use of systems engineering and requirements engineering is reinforced by typical modeling formalisms (Tolmer 2016; Figure 4.6). Indeed, the use of common modeling bases facilitates the continuity and consistency of conceptual models: this requirement is also presented in Figure 4.4.

If we start from the hypothesis that the IM (digital model) of a structure carries a specific value complementary to that of the physical structure itself, working on the organization of the enabling system (the project) appears to be more relevant today than managing the complexity of the system of interest (the work, the structure), which is often better controlled. The complexity of the works is generally lower or technically better controlled than that of the exchanges and information sharing, even for the most technically complex works. In projects, what today remains little or badly defined are not the technical specifications nor the technical requirements to be met but the requirements linked to the BIM approach, in particular from the point of view of information exchanges, collaborative work and steps for coordination. Today, digital technology questions practice more than technique (Dejoux and Léon 2018). Managing digital skills in companies and projects has become a real challenge due to the complexity and cross-functionality of the skills to be acquired, but also due to a lack of resources. The differentiation of the enabling system and the system of interest facilitates the management of these two project complexities (Figure 4.6). However, it has been shown that using these concepts and methodological tools only benefits the system of interest (the very reason for organizing a project) if they are mobilized by and for the enabling system (Figure 4.6). Indeed, requirements' verification takes place directly on the system of interest as part of accepting the construction works.

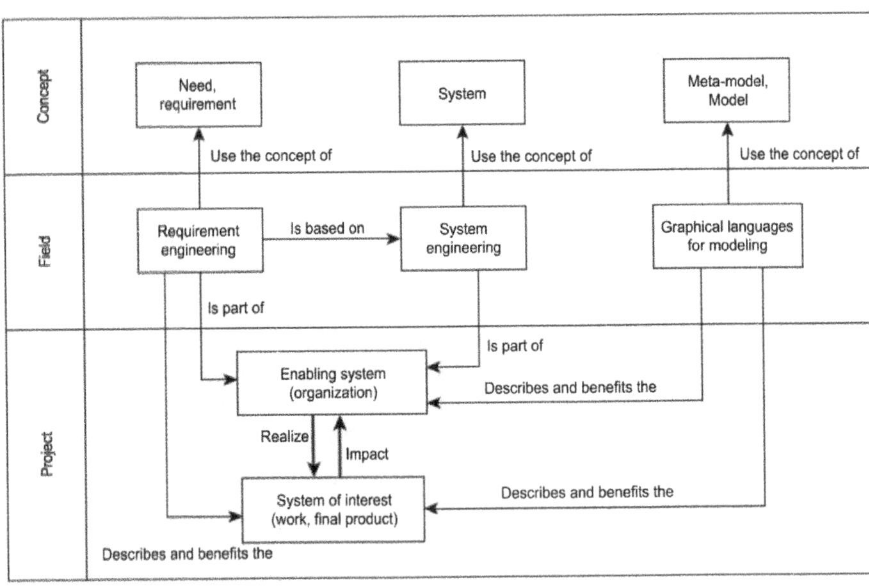

Figure 4.6. *Methodological concepts and tools for enabling system and system of interest (Tolmer 2016)*

Let us illustrate this point by what this implies for an invitation to the tender phase. For a BIM approach to benefit the work, it is necessary to apply a system engineering and requirements engineering approach to the enabling system. Here, we use the structured analysis and design technique (SADT) formalism. The BIM execution plan (the set of elements that describe the BIM-related part of the system of interest) is prepared by taking into account its environment (constraints, see Figure 4.7 and especially ISO 7817 (n.d.) about Level of Information Need tables) and the project inputs that relate to the work itself (the system of interest).

Tables 4.1 and 4.2 show in greater detail how the concepts of System Engineering and Requirements Engineering make it possible to structure the BIM processes of a project but also the more general processes (example here of the Project Management Plan, PMP), that is to say, the system of interest. Several types of requirements are considered here:

– products: performance, quality, business characteristics of objects or systems;

– process: regulatory procedures such as the water regulations, the building permit, validation circuits, etc.;

– modeling or information: the information that must be exchanged or modeled: 3D objects, properties of these objects, documents or other project data.

In a construction context, the system of interest is both the physical structure to be built and the information that describes it.

The current challenges of modeling are the durability, continuity, consistency and uniqueness of information over the entire lifecycle of a structure, an infrastructure, and the territory. In this sense, effective modeling of such a complex system, both from the conceptual point of view and the point of view of formalisms, must allow having sufficient proximity adapted to the reality of the work, as well as a level of detail respecting this reality about uses.

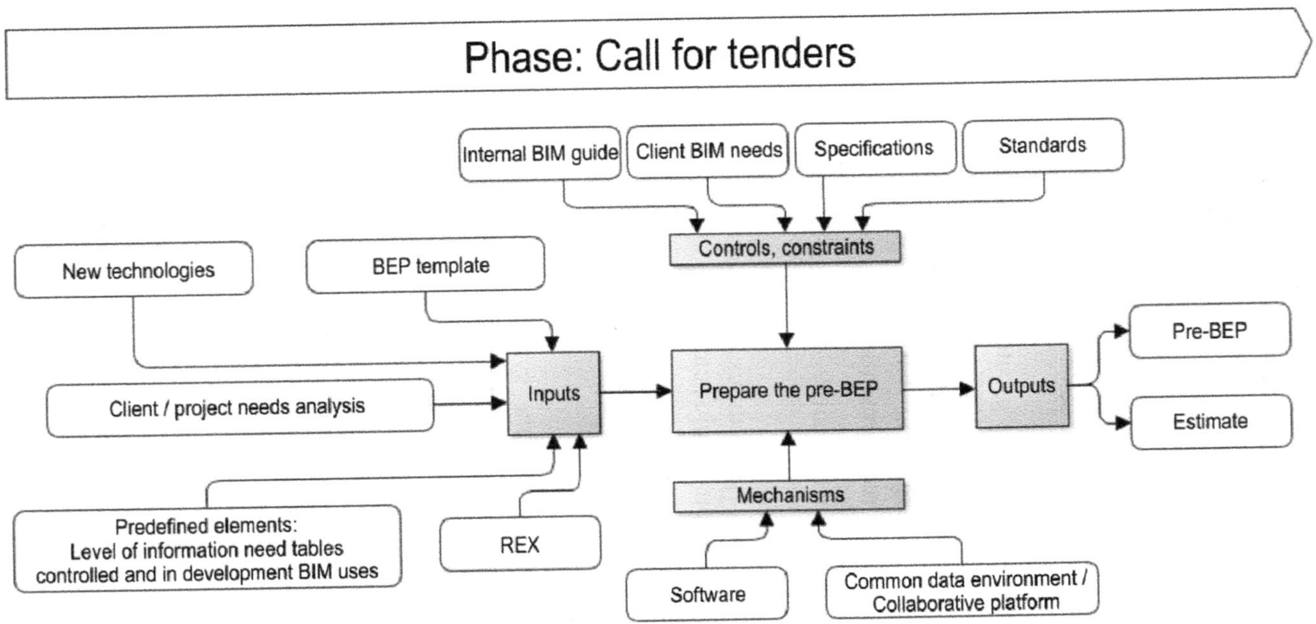

Figure 4.7. *SADT for the preparation of the BIM project agreement*

System engineering paradigm (A)		BIM Context (B)	
	(A) for (B)	Project management plan (PMP)	BIM execution plan (BEP) (ISO 19650-2:2018)
	System engineering:		
	System of interest	X	Constrains the BEP
	Enabling system	Described by PMP	Constrains the BEP (BIM uses)
	Requirement engineering		
	Product requirement	Described in project specifications	Are inputs to identify BIM uses
	Process requirement	Are included in PMP	Described in BEP
	Modeling / information requirement	X	Described in BEP

Table 4.1. *Interpretation and use of systems engineering concepts in a BIM context*

System engineering paradigm (A)		BIM Context (B)	
	(A) for (B)	BIM uses	Information needs (geometric and non-geometric)
	System engineering:		
	System of interest	Its quality is improved by following BIM uses	Is described by the information need
	Enabling system	Integrates BIM uses	Allows the definition, the production and the exchange of information for "system of interest" and "enabling system"
	Requirement engineering		
	Product requirement	Are inputs to identify the BIM uses	Used to identify the needed information to verify these product requirements
	Process requirement	Describe the BIM uses	Used to identify the information to be produced and exchanged
	Modeling / information requirement	Describe the information to be exchanged	Describe the information of each exchange or sharing of information

Table 4.2. *Interpretation and use of system engineering concepts in a BIM context ("BIM usage" and "information needed", as defined in EN 17412-1 (2020): Level of Information Needed)*

4.2.3. *The challenges associated with the modeling of complex systems*

While current practices in complex systems engineering rely on well-defined processes and innovative analytical approaches, practices around BIM must be seen through the lens of the many challenges related to integration and interdisciplinary needs. The degree of control over the different components of the complex system, that is, the digital twin, also varies. Indeed, while it is possible to control the technical or management subsystems, the independent components of a "system of systems" generate their own decisions, which are uncontrolled at a system level. The degree of control over these components depends on the ability of the overall system to control subsystems generating their own decisions.

Considering the three decompositions of systems engineering from ISO 19650-1 (2018) (Figure 4.5), augmented with the spatial decomposition necessary for construction works, decision-making cannot be based on the simple description of the built objects. Returning to the functional description is essential to visualize/represent the behavior of the digital twin. Depending on the event, it is necessary not only to identify which part of the physical structure requires an intervention, but also to describe the behavior of the associated structure. The users of these subsystems expect tools allowing them to preview and correct the states of the complex system, while taking into account its self-organization and adaptation to its environment.

	Explanation
Certifying knowledge	Enabling non-expert users to assess the degree of usefulness of expert knowledge in decision-making.
Assembling knowledge	Integrating knowledge from different sources to support the needs of the decision-maker.
Translating or disassembling knowledge	Converting complex or overly broad concepts into a framework for decision-making (possibilities, motivations).
Providing knowledge	To support decision-making.

Table 4.3. *Challenges associated with the management of expert knowledge in a decision-making process*

4.3. Conclusion

We are now moving into the abstract design paradigm. The convergence and continuity of modeling will facilitate managing of building information and knowledge of the structure. To achieve this, it is necessary to ensure a continuity of interpretation between numerous concepts and models of thought: requirements management, which is often considered the entry point for satisfying the client's needs, systems engineering, conceptual data models specific to the various trades and fields of activity, modeling formulas, exchange formats, etc. Indeed, the construction of the urban environment integrates all types of buildings, infrastructures, industrial works, the ground, and the subsoil, which all have different interpretations for previously mentioned concepts. From a low-carbon trajectory perspective, for example, the urban environment's design, construction and operation will require the use of data, information and knowledge that will necessitate such coherence within the digital twin.

4.4. References

Ackoff, R. (1989). From data to wisdom. *Journal of Applied Systems Analysis*, 16, 3–9.

Azhar, S. (2011). Building Information Modelling (BIM): Trends, benefits, risks and challenges for the AEC industry. *Leadership and Management in Engineering, ASCE*, 11, 241–252.

Bot, L. and Vitali, M.L. (2011). *Modélisation et activités des ingénieurs*. L'Harmattan, Paris.

Bunge, M. (1979). *Treatise on Basic Philosophy, in Ontology: A World of Systems*. D. Reidel, Dordrecht.

Déjoux, L. (2018). *Métamorphose des managers à l'ère du numérique et de l'intelligence artificielle*. Pearson, Montreuil.

Fiorèse, S. and Meinadier, J.-P. (2012). *Découvrir et comprendre l'ingénierie système*. Cépaduès édition, Toulouse.

ISO 7817 (n.d.). Building information modelling – Level of information need – Concepts and principles.

ISO/TC 59/SC 13 (n.d.). Organization and digitization of information about buildings and civil engineering works, including building information modelling (BIM).

Levin, M.S. (2006). *Composite Systems Decisions*. Springer, New York.

Micaelli, J.P., Bonjour, E., Deniaud, S., Loise, D. (2011). L'ingénierie système fondée sur les modèles. In *Modélisation et activités des ingénieurs*, Bot, L., Vitali, M.L. (eds). L'Harmattan, Paris.

MINnD (2019). S1 UC1 Th3, Approche holistique des infrastructures Cadre de référence – Parties 1 et 2.

Norme NF EN 17 412-1 (2020). Building Information Modelling – Level of Information Need – Part 1: Concepts and principles.

Norme NF EN ISO 14258 (1998). Industrial automation systems – Concepts and rules for enterprise models.

Norme ISO 772 (2011). Hydrometry – Vocabulary and symbols. International Organization for Standardization, Geneva.

Norme ISO 17261 (2012). Intelligent transport systems – Automatic vehicle and equipment identification – Intermodal goods transport architecture and terminology. International Organization for Standardization, Geneva.

Norme ISO 11354-1 (2011). Advanced automation technologies and their applications – Requirements for establishing manufacturing enterprise process interoperability – Part 1: Framework for enterprise interoperability. International Organization for Standardization, Geneva.

Norme ISO 19650-1 (2018). Organization and digitization of information about buildings and civil engineering works, including building information modelling (BIM) – Information management using building information modelling – Part 1: Concepts and principles. International Organization for Standardization, Geneva.

Norme ISO/IEC/IEEE 15288 (2015). Systems and software engineering – System life cycle processes. International Organization for Standardization, Geneva.

Rochet, S. (2007). Formalisation des processus de l'ingénierie système, proposition d'une méthode d'adaptation des processus génériques à différents contextes d'application. PhD Thesis, Université Paul Sabatier, Toulouse.

Rochet, C. and Peignot, J. (2013). De la gestion des villes à la conception d'écosystèmes urbains durables. In *MED6*. Marseille.

Tolmer, C.E. (2016). Contribution à la définition d'un modèle d'ingénierie concourante pour la mise en œuvre des projets d'infrastructures linéaires urbaines : prise en compte des interactions entre enjeux, acteurs, échelles et objets. PhD Thesis, Université Paris Est, Paris.

5

Complex Systems Modeling Approaches

Ana ROXIN[1] and Christophe CASTAING[2,3,4]

[1]Laboratoire d'Informatique de Bourgogne, Université Bourgogne-Franche-Comté, Dijon, France

[2]buildingSMART International, Paris, France

[3]Groupe Egis, Paris, France

[4]Projet MINnD, Paris, France

5.1. Introduction

In this chapter, we present two main approaches for modeling complex systems, namely approaches based on data models and knowledge models. For each approach type, we summarize their history underlying modeling principles and the international standards that use them. However, before going into detail and to better understand the differences between these approaches, we start with a brief overview of the reference framework for data governance, for example, the ISO 8000 series of standards. Published by ISO/TC 184/SC 4, this series deals with data quality principles and specifies characteristics for assessing the quality (or degree of compliance) of an organization's data. Data quality is a determining factor for the quality of information and, consequently, the accuracy and reliability of the knowledge that can be deduced from this information. Indeed, information is data that is placed in a context, and knowledge is information coupled with experience or know-how. According to ISO 8000-1 (ISO/DIS 8000-1 2011), data have the attributes of provenance, accuracy and completeness. They follow a formal syntax and use concepts defined in dictionaries.

Building Information Modeling,
coordinated by Régine TEULIER and Marie BAGIEU.

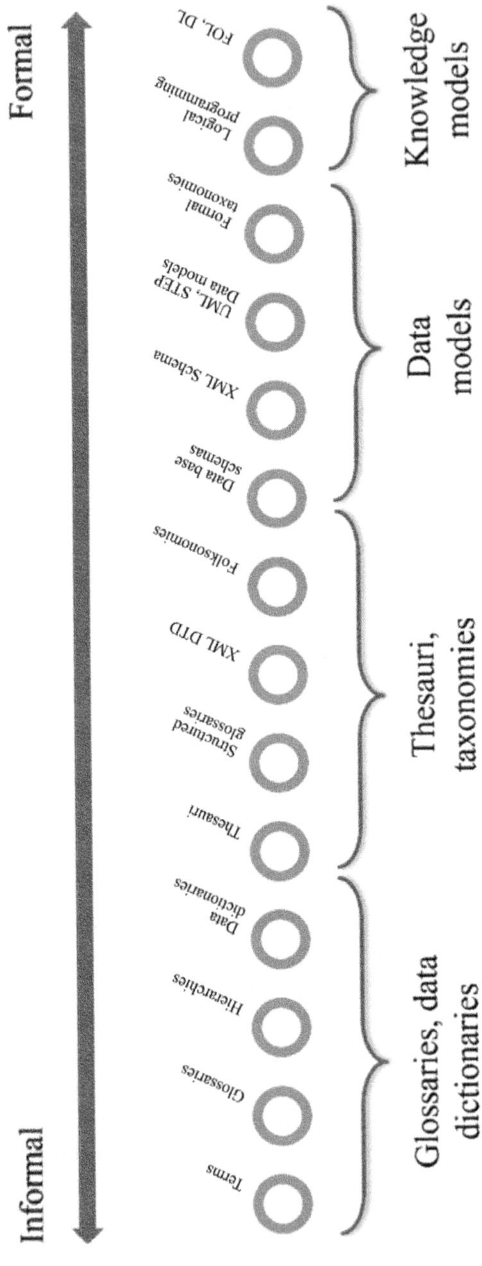

Figure 5.1. *Data structuring approaches according to the formality of the language used*

These elements are found in the existing computer approaches for data structuring. These approaches are implemented as computer languages and can be classified according to their degree of formality. In computer science, such a language is said to be formal and includes an alphabet (set of elements), rules to determine whether an element belongs to the alphabet of the language (grammar) and whether a set of elements respects these rules (syntax). In computer science, the elements of a formal language, in addition to syntax, also have a meaning or semantics. The more explicitly and logically semantics are defined, the more formal the language is. This is illustrated in Figure 5.1. Data dictionaries, in the computer sense, are considered informal because they contain only terms and their definitions. These terms (their semantics) can only be interpreted by a human user and not by a machine. The same can be said about glossaries and term hierarchies. On the opposite, logic languages, based on first-order logic, represent approaches that explicitly specify the meaning of statements, thus allowing their interpretation by a machine (through the execution of algorithms called decision procedures).

In the context of building information modeling (BIM), various standardization works within ISO provide a complementary framework to IT approaches for data dictionary design. We can mention here the following standards:

– ISO 12006-2 (ISO 12006-2 2015) is intended for organizations developing and publishing classification systems and tables. ISO 12006-2 (2015) specifies classification table headings for mutual harmonization;

– ISO 23386 (2020) specifies rules for defining and managing construction properties. A set of attributes to be used in defining such properties is described. Business process modeling notation (BPMN) processes are specified for digitally exchanging such digital (groups of) properties.

However, none of these standards specifies a type of model (in the sense of Figure 5.1) to be used for underlying structuring data. Yet, the need for model engineering goes hand in hand with the need for digital processing of data interpretations according to a given context (or, more precisely, with the need to control these interpretations).

From a computer science perspective, as illustrated in Figure 5.1, there are several families of models, notably data models and knowledge models. The languages for data models (e.g. unified modeling language [UML]) are more formal than those for defining a dictionary. Indeed, the rules of element composition (syntax) are more constrained. However, data meaning is not formally specified: data models (graphically) represent knowledge (for this reason, they are said to be significant); nevertheless, an algorithm cannot reason upon this knowledge. A human user, an expert in the field, must interpret the knowledge represented in this way.

Approaches based on logical languages constitute the family of knowledge models. Their degree of formality is maximal (more formal than data models), and they constitute explicit and formal knowledge specifications. Therefore, it is possible to use an algorithm for simulating human reasoning upon this knowledge. Moreover, such reasoning can be specified through logical languages, and all the algorithm's deductions can be explained logically (e.g. in the form of decision trees).

Sections 5.2.1 and 5.3.1 present the state of the art of model-based and knowledge-based approaches. Section 5.3.2 presents a comparative discussion of the two approaches from a computer science point of view. Sections 5.2.2 and 5.3.3 summarize the uses of these approaches in infrastructure construction projects from an industrial point of view and in terms of standardization.

5.2. Object model-based approaches

UML and the associated formalisms appeared in the more global context of object-oriented design. Indeed, the so-called object (or class) based models appeared in the early 1990s, in a context where data models of the {Entity/Relationship} type, as well as flow diagrams, were already widely used by companies for system requirements (as early as the 1970s). To differentiate existing modeling approaches, a distinction is made according to what the modeling is about: the nature (or type) of an activity (or a business domain – problem modeling; or a system that implements it – solution modeling).

Therefore, when the term object-oriented analysis was coined by Coad et al. (1990), it referred to an approach for integrating services and messages (concepts from object-oriented programming) into {Entity/Relationship} models. The initial idea was to improve the inheritance management done by the latter. For Coad and Yourdon, an "object" was "an abstraction of something in the problem space, reflecting the capabilities of a system to hold information about it, interact with it, or both; an encapsulation of attribute values and their exclusive services" (Coad et al. 1990). According to this definition, an object-oriented analysis is aimed at the problem space (e.g. the company, the organization or the business domain under consideration) and not at the solution space (e.g. computers, languages and computer programs). Therefore, modeling with an object-oriented approach is initially aimed at the problem space and not the solution space. Unfortunately, this link with the problem space has become weaker and weaker. Indeed, when UML appeared, the analysis aspects of the non-computer environment seemed to have been lost.

This was confirmed in 1999, with the object definition provided by Rumbaugh et al. (1999). An object is defined as "an instance of a class". Still, according to Rumbaugh et al., a class is a "descriptor for a set of objects that share the same attributes, operations, methods, relations and behaviors" (1999). Everything is an object with these definitions, which are the law in the UML community. The link with the (computer) implementation of the system becomes intrinsic to all UML modeling. A UML model thus emphasizes object orientation rather than the analysis of a problem space. We shall see in section 5.3 that knowledge-based models are more suitable for problem modeling. The UML language is central to model-based approaches (MDAs, or model-driven architecture). The following section presents these MDA approaches and their layered structure.

5.2.1. *Model-based architectures and standards*

Specified in the framework of the object modeling group (OMG) and taking up the principles of model-driven engineering (MDE), model-based architectures (model-driven architecture or MDA) define rules for structuring system representation as models. MDA approaches (Miller et al. 2003) consider three abstraction levels from which a system can be described: CIM, PIM and PSM. In the MDA approach, such a level is considered "an abstraction technique that allows one to focus on a particular set of problems within a system while removing all irrelevant details. A level of abstraction can be represented via one or more models" (Tuyen 2006). The three abstraction levels correspond to three model types:

– The computation independent model (CIM) defines the system's context and requirements (prerequisites), without taking into account its structure or processes. The UML (Booch et al. 1998) is used to define this type of model (ISO/IEC 19505 2012).

– The platform-independent model (PIM) describes the operational capabilities of a system (defined as abstractions of the platform) without taking into account implementation details specific to the platform (or to a set of platforms).

– The platform-specific model (PSM) represents a translation of a PIM concerning a specific platform. PSMs are defined through so-called implementation languages (e.g. Java, Python or XML Schema). Automated tools (e.g. model transformation tools) translate the PIM into different PSMs.

Furthermore, MDA approaches are based on a so-called metamodel architecture structured in layers. These layers (illustrated in Figure 5.2) contain models that can be described from a different point of view. A model of the M1 layer can describe a CIM abstraction, a PIM view or a PSM implementation.

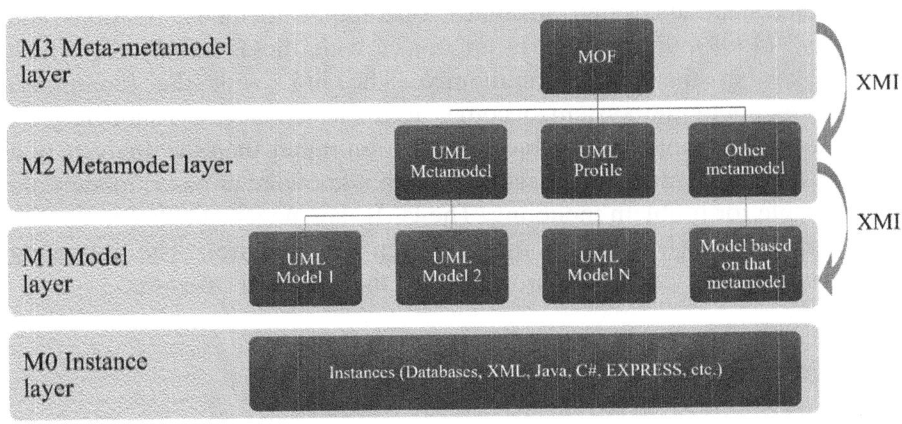

Figure 5.2. *Layers of the MDA*

The four layers shown in Figure 5.2 are defined in the following.

1) The meta-metamodel layer (M3) is defined through the meta-object facility (MOF) standard (2014). The latter defines an abstract language and a framework for the specification, construction and management of metamodels independently from a technology. The MOF language allows it to represent itself; it is said to be self-describing. It can also represent all the other languages of the MDA approach (e.g. UML).

2) The metamodel layer (M2) includes languages specified according to the MOF standard, including UML (Booch et al. 1998) and XMI (XML metadata interchange) (XMI 2014). The UML language allows specifying

different aspects of a system (whether it is an organization or software), notably its static aspects (e.g. class diagrams, component diagrams), dynamic aspects (e.g. runtime behavior, business processes through activity or sequence diagrams), as well as the relationships between users and the system's functionalities (e.g. use case diagrams). The XMI standard enables the exchange of models represented as XML documents. The EML (Eclipse Modeling Framework) standard allows models to be represented as Java classes while allowing operations to be performed on these models, for example, validation or transformation (Budinsky et al. 2003). The SysML language (OMG 2002) is another example of a metamodel or M2-level language. Developed as a UML 2 profile, SysML is a graphical representation language widely used in systems engineering within the ISO/TC184/SC4 Industrial Data framework.

3) The model layer (M1) contains user-made models, often in the form of classes.

4) The instance layer (M0) contains the instances of the models of the upper layer (M1). These instances generally correspond to the elements of the real world, modelled through the models of the M1 layer.

Other approaches from the MDA family are relevant for modeling complex systems. We can mention the model-based system engineering (MBSE) approach, developed under the International Council on Systems Engineering (INCOSE) framework. This approach supports the specification of systems, requirements, design, analysis, verification and validation. It applies to all system lifecycle phases (Long et al. 2011). Related to MBSE is the model-based requirements engineering (MBRE) approach, which concerns the engineering and management of requirements based on models. The MBRE approach is independent of any IT tool and can therefore be implemented using any tool or combination of tools (Holt et al. 2012). The MBRE approach relies on graphical elements to exchange more precise and concise requirements. With this approach, requirements are considered one at a time, relative to their definitions, and then assigned meaning by placing them in the appropriate context. A requirement in a context (the actors considered or the system's hierarchical levels) is called a use case.

5.2.2. *International standards using this type of modeling*

The MDA approach is used in geographic information systems (GIS). The associated standards are based on different schemas, models and metamodels whose layer structure corresponds to the MDA approach. According to the MDA approach, the language used for defining GIS conceptual models is the UML language, as standardized by ISO 19505-2 (ISO/IEC 19505-2 2012). GIS models are structured according to the four layers of the MDA approach. Different international standards have been defined for each of these layers, listed in Table 5.1.

	Standards
M3 metamodel layer	Meta-Object Facility (MOF) ISO/IEC 19508 (2014)
M2 metamodel layer	UML Metamodel ISO/IEC 19505 UML core profile ISO 19103 UML profiles for ISO 19103 GIS application diagrams
M1 model layer	Application schemas: INSPIRE GML, OGC CityGML, OGC InfraGML, OGC IndoorGML, OGC GeoSciML
M0 instances layer	XML, XSD – XMI (ISO/IEC 19509) Java – JMI (Java Metadata Interchange)

Table 5.1. *International standards using object-based modeling*

The INSPIRE (INfrastructure for Spatial InfoRmation in the European community) 2007/2/EC European directive also builds upon MDA principles to specify principles for interoperability and accessibility of geographic information. In the INSPIRE sense, "interoperability" represents compatibility between two systems that allows them to exchange information so that other systems can understand them (Ansorge et al. 2016). The INSPIRE framework specifies system functionality as a PIM. This PIM is a data specification for all INSPIRE-compatible systems and represents a resolution of the conceptual interoperability between systems. The language used to define this PIM is UML. The INSPIRE directive considers translating the PIM into PSMs for operational interoperability. These translations are performed via automatic procedures based on languages such as Java, XML Schema or Python (Ansorge et al. 2016). The specification of the INSPIRE conceptual model is available in INSPIRE (2013).

Table 5.1 allows for introducing Figure 5.3, which depicts the four MDA levels concerning their use in Industry Foundation Classes (IFC) (ISO 16739-1 2018).

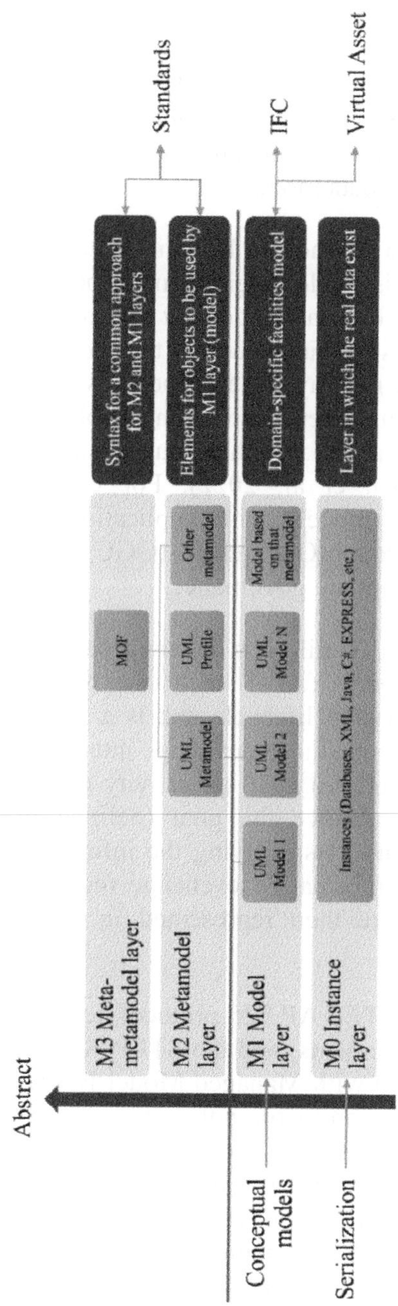

Figure 5.3. *Interpretating the MDA approach in the context of MINnD*

The STEP standard (ISO 10303 1994) defines the infrastructure construction or engineering projects framework. The ISO 10303 family of standards defines an approach for representing and exchanging product data mainly based on the SysML language (OMG 2002). It provides a neutral mechanism for describing products throughout their lifecycle. The ISO 16739 IFC (ISO 16739-1 2018) standard represents an equivalent derived from STEP for construction data models.

As part of the works to extend the IFC model to infrastructures, it was necessary to understand this model engineering approach to determine how to complete the initial conceptual model: by appropriating the metamodel underlying the IFC4 version, introducing the domain-specific business models such as sanitation, railway ballast, energy supply, as UML models, and then producing the implementation in the application schema (PSM). The correspondence with Table 5.1 is then found, allowing for a new step in a model federation (Roxin et al. 2022). For example, geotechnics is developing PIM business models with application schemas specific to geographic information through GeoS ciML (OGC 2017) and BIM through IFC, accompanied by appropriate governance.

The ISO 10303 standard family has different parts corresponding to the so-called application protocols (APs), each of which defines information exchanges throughout a system lifecycle and is a response to a set of use cases and business requirements, enabling interoperability of product information. Each AP includes a goal, an activity diagram describing what an engineer must do to achieve the goal (defined in BPMN), and an application requirements model specifying the information needs related to the engineer's activities (information exchange requirement [IER]). These needs and requirements are then represented in the common integrated resources (IR) set.

In a BIM context, the STEP AP 225 protocol describes the requirements associated with building elements, and the STEP AP 242 protocol specifies the requirements associated with Managed Model-Based 3D Engineering. It should be noted that this protocol, not used in construction, could be integrated with the previous ones in the context of the digital twin development, similarly as GIS-related protocols (mentioned in Table 5.1). According to these protocols, and still, in a BIM context, needs and requirements are specified in natural language, according to the methodology defined by the Information Delivery Manual (IDM) standard (ISO 29481-1 2016). According to the IDM standard, the formal syntax to be used for exchanging construction data is the one specified by the IFC (ISO 1639-1 2018): it is either the EXPRESS format, as defined in (ISO 10303-21 2016), or the XML format.

Still in a BIM context, among these application protocols, we can also mention the AP233 "System Engineering Data Representation", whose concepts, which correspond to the functional and structural breakdown (ISO/TS 10303-1216 2008), have been standardized by the Product Lifecycle Support (PLCS) project (Eckert et al. 2005).

In the context of the development of the digital model of the territory, new needs arise to manage, on the one hand, the continuity of data exchanges throughout a structure's lifecycle and, on the other hand, the implementation of standards as a contractual basis for defining information. The governance of these standards must include a quality process that verifies that business practices are reflected in the different stages of formalization: a formalism adapted to the different requirements. The two modeling approaches presented in this chapter make it possible to complete the formalism tools. However, to verify and validate the correct relationship between the business requirements, the platform requirements and the correct implementation in the software, it is necessary to accompany this formalism with clear and detailed documentation.

5.3. Knowledge model-based approaches

As stated in this chapter's introduction (and illustrated in Figure 5.1), knowledge-based models are explicit and formal specifications of knowledge using logical languages can be interpreted by algorithms. Compared to the data models seen previously, knowledge-based models make it possible to specify problems in the context of a business domain without any link to a computer implementation. This is called semantic modeling. Knowledge-based models use the concept of class differently than object-oriented models:

– in semantic modeling, an entity is not concerned with operations, methods or behavior (contrary to an object in the object-oriented world). These elements belong to the domain of "process modeling";

– in a semantic model, the class of an entity is not simply a class of the set of "discrete entities with well-defined boundaries and identities" (as is the case for UML). A class here is limited to what Richard Barker calls classes of "things or objects of significance, whether real or imagined, about which information needs to be known or held" (Baker 1990, author's translation). We can summarize this by saying that in semantic modeling, a class groups together instances with the same interpretation (the same meaning), which share a set of properties.

Different types of knowledge models can be defined depending on the family of logical languages used. To characterize them, two indicators are essential: the expressiveness and the decidability of the language used. To simplify, we can define the expressiveness of a formal language as the number of operators or constructors that can be used to combine concepts to form new ones (according to language syntax rules). Decidability can be defined as the ability to apply deduction algorithms to models described with the logical language being considered. A deduction algorithm must be able to verify in a finite time that all statements respect both the syntax of the language (correctness of the model) and the associated semantics (completeness of the model). The more expressive a logical language is, the less decidable it is. In other words, the more logical operators and constructors a language uses, the less it is possible to verify the completeness and correctness of models defined with this language in a finite time. For the sake of implementation efficiency, it is appropriate to find a language that achieves this balance between expressiveness and decidability (Roxin 2018). It is the case for Description Logics (DL) languages, which specify ontologies. We present ontology-based approaches, manipulated elements and related computer standards in what follows.

5.3.1. *Presentation of the approach and associated standards*

Similar to object-based models, an ontology includes classes, properties and instances. However, an ontology allows properties to be defined independently of classes. Properties are used to specify the conditions that an instance must fulfill to belong to a class. These are either necessary conditions (instance membership to a class implies the instance respecting these conditions) or necessary and sufficient conditions (instance membership to a class is equivalent to the instance respecting these conditions).

An ontology distinguishes between assertional knowledge (or instances corresponding to the M0 level in the MDA approach) and terminological knowledge (related to layers from M1 to M3 in the MDA approach). An ontology necessarily contains terminological knowledge (or TBox). If an ontology also contains assertional knowledge (or ABox), it is called a knowledge base. An algorithm called a reasoner can be executed on an ontology or knowledge base. The verifications and deductions associated with each type of knowledge are illustrated in Figure 5.4. Such an algorithm makes it possible to deduce and materialize (in the ontology or the knowledge base) an ensemble of implicit knowledge (not initially explicit). This process is called inference, which is why the algorithm that implements it is called an inference engine. Finally, when implementing an approach based on such a knowledge base, a graphical interface is often added to facilitate interaction with the knowledge base and the inference engine (see Figure 5.4).

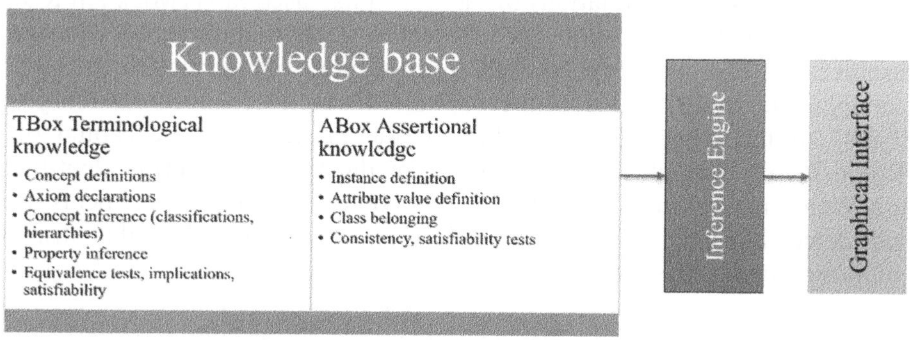

Figure 5.4. *The general architecture of knowledge model-based approaches*

The computer languages used to specify and implement ontologies are defined and standardized by the W3C (World Wide Web Consortium). All these languages and technologies constitute the so-called Semantic Web technologies for which the W3C has defined a layered architecture (see Table 5.2). We present it and summarize the associated languages in the following paragraphs.

	Standards
Model layer	RDF Schema, OWL, SKOS
Information exchange layer	RDF (Resource Description Format)
Identification and encoding layer	URI (Uniform Resource Identifier), Unicode

Table 5.2. *International standards for knowledge modeling*

To better understand the Semantic Web, we begin by introducing the principles of Linked Data. The Linked Data approach represents a subset of the principles and technologies of the Semantic Web, aiming at sharing and reusing data at the Web scale. Indeed, it is a matter of taking the principles of the current Web architecture and extending them to describe knowledge (see Table 5.2). As with the classic Web, the Semantic Web relies on using URIs as a unique and global identification mechanism for resources at the Web scale. The HTTP protocol is the universal access mechanism to resources, as it is for the traditional Web. The difference lies in the description of the resources. While the traditional Web uses HTML for web pages, the Semantic Web relies on the RDF model (Hayes et al. 2014), which is based on a graph structure and allows for the use of typed links (thus overcoming the limitations of <a href> links used in HTML). An RDF model is a directed, labeled graph of resources identified through URIs. The resources are in triples respecting the form <Subject Predicate Object>. For example, the triplet <Paris is_the_capital_ofFrance> connects the subject "Paris" to the object "France" through the property "is_the_capital_of" (Roxin 2018). Each element (the subject, predicate, and object) is identified at a Web-scale by an HTTP URI. In this context (Web), two elements are considered identical by a reasoner if their identifiers (thus the strings composing their URIs) are identical (Roxin 2018).

	Traditional Web	Semantic Web
Description	HTML	RDF
Access	http (HyperText Transfer Protocol)	
Identification	URI (Uniform Resource Identifier)	

Table 5.3. *Classic Web principles compared to Linked Data principles*

Following the identification and access principles of Table 5.3, all elements of RDF, ontology description languages and any ontology defined with these languages are published on the Web and identified with HTTP URIs. It is possible to avoid writing long strings using the URI abbreviation schema called *qname*. In its simplest form, a URI expressed according to this schema consists of two parts: a domain name and an identifier, separated by ":". Therefore, the prefix "rdf" corresponds to the domain name where the RDF model is published, that is, the string "http://www.w3.org/1999/02/22-rdf-syntax-ns#". Therefore, rdf:Resource represents an abbreviation of the complete HTTP URI allowing the identification of the "Resource" concept as defined in RDF (http://www.w3.org/1999/02/22-rdf-syntax-ns#Resource).

Ontology description languages complement these principles and involve using the RDF model. Indeed, RDF is the generic model for describing resources, notably through its highest level concept rdf:Resource. In RDF, everything is a resource: RDF does not allow distinguishing between an instance and a class. This abstract functionality of RDF makes it the ideal candidate for integrating heterogeneous data. However, when it comes to describing knowledge in a business domain, additional terms are needed. They are specified through RDF Schema and Web Ontology Language (OWL).

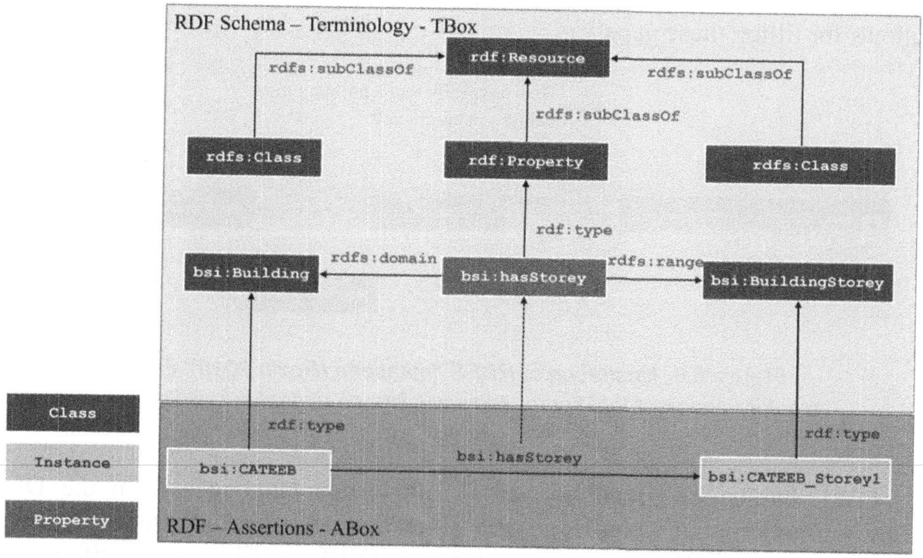

Figure 5.5. *Illustration of the new concepts brought by the RDFS language compared to RDF (Roxin 2018). For a color version of this figure, see www.iste.co.uk/teulier/building.zip*

RDF Schema (Brickley et al. 2014) introduces a vocabulary on top of RDF, with different terms to further characterize statements formed with RDF. The rdfs prefix is used for identifying RDF Schema terms; it corresponds to the domain name "http://www.w3.org/2000/01/rdf-schema#". With RDF Schema, it is possible to define classes (rdfs:Class), subclasses (rdfs:subClassOf) and subproperties (rdfs:subPropertyOf) as well as property domains (rdfs:domain) and ranges (rdfs:range). Figure 5.5 illustrates the new terms brought by RDF Schema compared to RDF. Therefore, an instance can belong to a class (via the rdf:type property). A class and a property remain subclasses of the abstract concept rdf:Resource.

Despite its contributions, RDF Schema has limitations that do not make it a suitable candidate for modeling complex knowledge. Let us consider the model described in Figure 5.6, a building which has floors that can be of two types: either "basement" (bsi:Basement) or "floor" (bsi:Floor). To specify the class of a building without a basement (bsi:BuildingWithout Basement), one would need to be able to specify a restriction on the bsi:hasStorey property (must not contain a basement type floor). RDF Schema does not allow adding this kind of constraint. The OWL language family allows for filling these gaps.

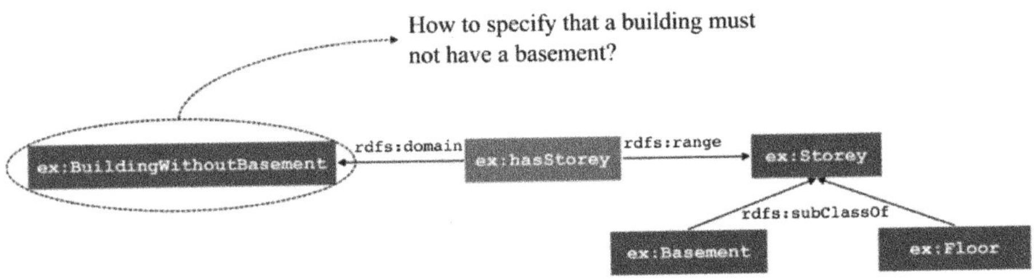

Figure 5.6. *Illustration of RDFS limitations (Roxin 2018). For a color version of this figure, see www.iste.co.uk/teulier/building.zip*

OWL languages (Bechhofer et al. 2004) use languages from the DL family to axiomatize knowledge of a given business domain. Depending on the implemented DL constructors, different OWL profiles are specified by the W3C. Their study is beyond the scope of this chapter. The prefix owl is used to identify OWL terms; it corresponds to the domain name "http://www.w3.org/2002/07/owl#". The OWL language family defines two types of properties: (a) object properties (owl:ObjectProperty) and (b) datatype properties (owl:DatatypeProperty). The latter are used in triples with a value or a datatype as their object, whereas object properties are used in triples with classes or instances of classes as their object. OWL also provides properties for specifying "synonymy links" between classes, properties and instances (owl:equivalentClass, owl:equivalentProperty and owl:sameAs, respectively). OWL also allows defining restrictions on property values (owl:allValuesFrom, owl:someValuesFrom, owl:hasValue) as well as cardinalities (owl:minCardinality, owl:maxCardinality, owl:cardinality).

To query knowledge models specified with RDF Schema or OWL, the W3C has specified the SPARQL Protocol and RDF Query Language (Harris et al. 2013). SPARQL includes a protocol for addressing queries via http and a language specification for composing such queries. SPARQL represents the equivalent of Structured Query Language (SQL) for knowledge models.

5.3.2. *Discussion*

Semantic technologies provide a generic and flexible means for discovering and integrating data from distributed data sources on the Web. Compared to object-based modeling approaches (e.g. APIs, relational databases), knowledge-based approaches (semantic modeling) have several advantages, which are summarized below.

Traditionally, object-oriented modeling starts with defining a structure (schema), which the user then instantiates with actual data (from the Web, applications, databases, etc.). With semantic approaches, it is still possible to work this way, and importantly, it is possible to reverse this process. It is possible to start directly from the data (regardless of their size or level of heterogeneity) and manually or automatically assign them to classes (either defined beforehand or generated from the data itself if coupled with machine learning approaches).

The RDF data model is an abstract language that can define data instances, data structures and their relationships. In other approaches, two languages would have been necessary to achieve this. For example, in the STEP approach (ISO 10303 1994), the EXPRESS language is used for data structures and the STEP Physical File Format (SPFF) language for data instances (ISO 10303-21 2016). RDF allows for the combination of multi-level declarations through its highest level concept (rdf:Resource), which can be defined at any meta-level.

In MDA approaches (including STEP technologies such as EXPRESS/SPFF/SDAI), a basic meta-concept is defined to which other concepts are linked. For example, in EXPRESS, the basic meta-concept is the Entity concept, which implements attributes and constraints. The definition of an attribute is always done in the context of an Entity concept. It is not the case with semantic approaches, where the classes and properties of an ontology are independent of each other. It is, therefore, possible to specify properties and constraints outside any class and to associate them with different classes.

In semantic modeling, the starting assumption is that the absence of information in an ontology is never interpreted as negative (open-world assumption). This is called Open World Modeling and is opposed to Closed World Modeling, which makes the opposite assumption (closed-world assumption) and characterizes object-oriented approaches. With the open-world assumption, if we specify that a building contains one floor, then run the query "how many floors does the building have", the reasoner cannot answer. It is because the building may contain other floors, and this knowledge has not yet been specified in the ontology. If we wish to have an answer to the query, we would have to specify that the building contains only one floor. The Open World Assumption is a direct consequence of the idea behind the Web, "anyone can say anything about anything", but it is also related to human knowledge being, by definition, incomplete.

Ontologies defined using standard Semantic Web languages are designed to be specified, extended and maintained progressively or incrementally. This is indeed a direct consequence of the closed-world assumption. At the start of the ontology modeling process, the ontology is empty, so a priori, everything is possible. Constraints are added during the (iterative) modeling process, which makes the ontology more restrictive. When we work under the open-world assumption, we have to specify what is not possible, forbidden, or excluded. This corresponds to how humans function in the real world: we are used to dealing with incomplete information. In contrast to open-world systems, one specifies explicitly what is possible in a closed-world system. In such a system, everything that cannot be deduced (or proven true) from the specified knowledge is considered false. This is notably the case with object-oriented approaches.

5.3.3. *International standards using this type of modeling*

The ISO/DIS 21597-1 (2020) Information Container for Data Drop (ICDD) standard specifies an ontology for a container for grouping documents and parts of documents (ISO 21597-1 2020). The first part of this standard is accompanied by a set of methods that use ontologies to link otherwise disjointed data within these documents. The second part extends the types of links that can be specified between documents (ISO 21597-2 2020).

This standard was designed to address the need in the construction industry for a uniform approach to organizing information for data exchange by providing tools for creating semantic links between concepts in separate documents.

Indeed, in the construction sector, information deliveries often consist of plans, information models (representing built or natural elements of the physical environment), text documents, spreadsheets, etc.

The ability to specify relationships between information elements contained in different documents can increase the value of such data exchange. The composition of such a container derives from both the process requirements (e.g. delivery of as-built information) and the specific functional objective (e.g. execution of a quantity takeoff or communication of aspects related to 3D models).

However, parts 1 and 2 of this standard have many limitations, mainly because the ontologies defined for the container and the linkset between documents were built from UML models. These ontologies are, therefore, unnecessarily complex, and industry implementations have demonstrated their limitations (especially in terms of reasoning). As given, their use is not very useful, let alone intuitive. However, part 2 of the standard specifies that it is possible to modify, extend or adapt the container and link ontologies.

This represents an opportunity not only to use ontology modeling languages (e.g. OWL) correctly to increase implementation efficiency but also to add additional functionality to adapt the container to a given purpose. Simplifying underlying ontologies can facilitate innovative software development while remaining compliant with the standard.

5.4. Hybrid approaches

Several researchers have investigated the adaptation of MDA approaches into ontology-based modeling approaches. It gave birth to the model-driven development (MDD) research branch. Djuric et al. (2005) define an ontology architecture based on the MDA approach. This architecture includes the specification of an ontology metamodel and a UML profile for ontologies. A transformation of the UML ontology into OWL has been implemented. The general approach is illustrated in Figure 5.7.

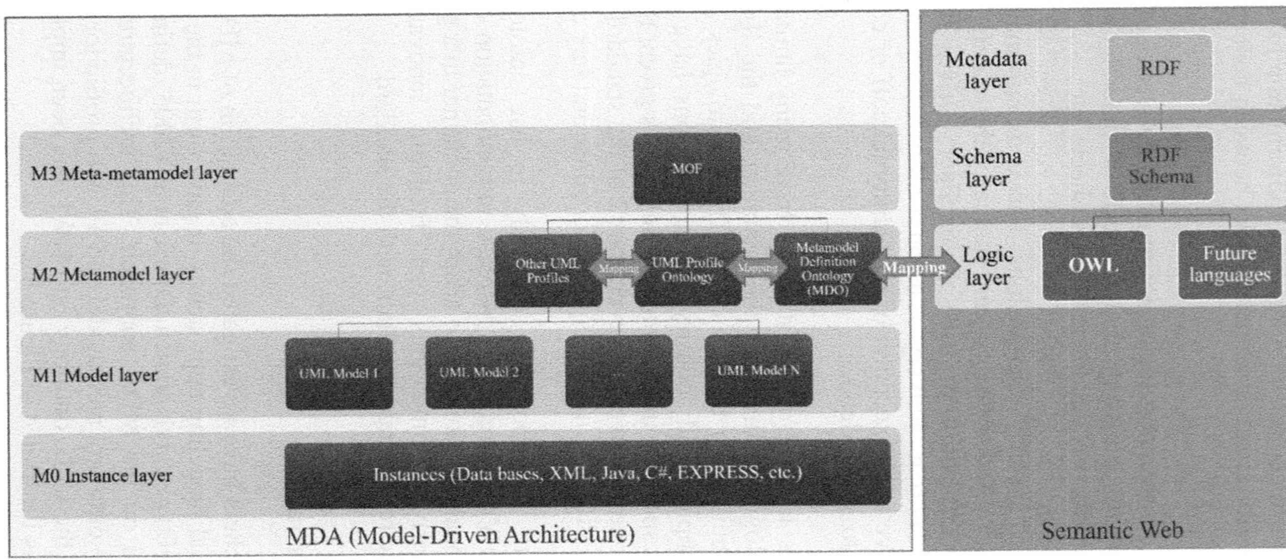

Figure 5.7. *Illustration of potential correspondences between MDA and OWL, adapted from (Djuric et al. 2005)*

ISO 19150-2 (2015) defines a set of rules for translating UML to OWL (ISO 19150-2 2015). However, various implementations of these translations have proven to have limitations. The resulting OWL models do not allow for efficient implementation and demonstrate poor performance from a reasoning perspective. This is mainly due to the conceptual differences between UML and OWL modeling, as summarized in the previous paragraphs. We can add that, in UML, a class is always the union of its subclasses, whereas this is not true in OWL. However, when OWL ontologies are generated from UML representations, OWL classes are defined as the union of their subclasses (owl:unionOf). This type of axiom negatively impacts the performance of reasoners, not to mention that it goes against OWL design principles. Unfortunately, this approach (automatic translation from UML to OWL) is increasingly employed.

However, some approaches correctly apply horizontal transformations between these two modeling approaches. We can cite the approach specified in the INSPIRE Directive (Ansorge et al. 2016). OWL can be considered for the specification of metamodels and conceptual schemas. UML can be used to represent these same elements. However, an alignment between two conceptual schemas is to be defined from OWL to UML, not vice versa.

Finally, the STEP architecture (ISO 10303 1994) translates SysML concepts into OWL to express the STEP conceptual model in OWL. Ontologies are also provided for defining dictionaries and other resources integrated in a BIM project. A detailed study of these approaches is beyond the scope of this chapter.

5.5. Conclusion

Despite various approaches to simulating human intelligence, computers have difficulty managing knowledge like humans. This is because knowledge is information interpreted concerning a context and an experience. In different situations, a human interprets two pieces of information differently. For a computer to perform a similar interpretation, it is necessary to specify an explicit model in a formal language.

In this chapter, we have presented two main approaches for modeling complex systems: approaches based on object models and approaches based on knowledge models. These two approaches come from different currents of computer science and allow us to answer different problems, problems that must be well-defined to be able to use the adapted approach. Approaches based on object models have a strong link with the computer implementation of the system considered. They are called semi-formal because they are a (graphical) representation of a (computer) implementation of a complex system.

The knowledge (i.e. the interpretation associated with the modeled information) is not explicitly modeled; it is implicit. A human expert is necessary to interpret such an object model correctly. This is not the case with knowledge-based models: they rely on formal logical languages to specify knowledge explicitly. Knowledge-based models are defined without any link to computer implementations. A computer can interpret these models (similar to a human user), and implicit knowledge can be inferred using reasoning algorithms.

Due to the differences between these two approaches, if transformations are to be considered from one to the other, it is critical to apply them in the right direction, from formal (knowledge-based) models to semi-formal (object-based) models. However, these transformations should be used sparingly and concerning a well-defined use case or problem. The two modeling approaches correspond to completely different conceptions, so an OWL model is not just a translation of a UML model into another file format and vice versa. Efficient use of the resulting models can be achieved by being aware of these differences and by using the modeling approach adapted to the targeted problem.

Even if knowledge-based models allow the interpretation of information to be specified in a specific context, it is still necessary to be able to define links between the interpretations in different contexts. In the domain of digital twins, this is all the more crucial as it is necessary to manage different abstractions (or views) of the knowledge of the structure (or building) at different moments of its lifecycle. It is necessary to be able to switch from one to the other and to ensure a continuum of interpretation between actors for each abstraction considered. This is only possible if semantic links (such as synonymy links, for example) are defined between the elements of the underlying knowledge models. The construction of the associated object models will ensure the adapted computer implementation of this interpretation continuum.

5.6. References

Ansorge, C., Borrebæk, M., Janssen, P., Van den Brink, L., Morrone, S., Parslow, P., Wilson, D., Robbrecht, J., Schleidt, K., Soucek, P. et al. (2016). The INSPIRE model-driven approach. INSPIRE Data Specification Extensions.

Baker, R. (1990). *CASE*Method: Entity Relationship Modelling*. Addison-Wesley, Wokingham.

Bechhofer, S., Van Harmelen, F., Hendler, J., Horrocks, I., McGuinness, D., Patel-Scneider, P., Stein, L.A. (2004). OWL Web Ontology Language Reference. *W3C Recommendation*, 10 February. [Online] Available at: https://www.w3.org/TR/owl-ref/ [Accessed 14 February 2023].

Booch, G., Rumbaugh, J., Jacobson, I. (1998). *The Unified Modelling Language User Guide*. Addison-Wesley, Wokingham.

Brickley, D. and Guha, R.V. (2014). RDF Schema 1.1. *W3C Recommendation*, 25 February. [Online] Available at: https://www.w3.org/TR/rdf-schema/ [Accessed 14th February 2023].

Coad, P. and Yourdon, E. (1990). *Object-Oriented Analysis*. Yourdon Press, Englewood Cliffs.

Djurić, D., Gaševic, D., Devedžic, V., Damjanovic, V. (2005a). A UML profile for OWL ontologies. Model driven architecture (MDAFA), lecture notes. *Computer Science*, 3599, 204–219.

Djurić, D., Gašević, D., Devedžić, V. (2005b). Ontology modelling and MDA. *Journal of Object Technology*, 4(1), 109–128.

Eckert, R., Mansel, W., Specht, G. (2005). Model transfer among CASE tools. *System Engineering*, 8(1), 41–50.

Harris, S., Seaborne, A., Prud'hommeaux, E. (2013). SPARQL 1.1 query language [Online]. Available at: https://www.w3.org/TR/sparql11-query/ [Accessed 14 February 2023].

Hayes, P.J. and Patel-Schneider, P.F. (2014). RDF 1.1 semantics [Online]. Available at: https://www.w3.org/TR/rdf11-mt/ [Accessed 14 February 2023].

Holt, J., Perry, S., Brownsword, M., Cancila, D., Hallerstade, S., Hansen, F.O. (2012). Model-based requirements engineering for system of systems. In *Proceedings of the 7th International Conference on System of Systems Engineering (SoSE)*, Genoa [Online]. Available at: https://doi.org/10.1109/SYSoSE.2012.6384145 [Accessed 14 February 2023].

INSPIRE (2013). INSPIRE Generic Conceptual Model. D2.5: Generic Conceptual Model [Online]. Available at: https://inspire.ec.europa.eu/documents/Data_Specifications/D2.5_v3.2.pdf [Accessed 14 February 2023].

ISO 10303 (1994). Industrial automation systems and integration – Product data representation and exchange. International Organization for Standardization, Geneva.

ISO 10303-21 (2016). Industrial automation systems and integration – Product data representation and exchange – Part 21: Implementation methods: Clear text encoding of the exchange structure. ISO/TC 184/SC 4 Industrial Data. International Organization for Standardization, Geneva.

ISO 16739 (2018). Industry Foundation Classes (IFC) for data sharing in the construction and facility management industries. International Organization for Standardization, Geneva.

ISO 16739-1 (2018). Industry Foundation Classes (IFC) for data sharing in the construction and facility management industries – Part 1: Data schema. ISO/TC 59/SC 13 Organization and digitization of information about buildings and civil engineering works, including building information modelling (BIM). International Organization for Standardization, Geneva.

ISO 19150-2 (2015). Geographic information – Ontology – Part 2: Rules for developing ontologies in the Web Ontology Language (OWL). ISO/TC 211 Geographic information/Geomatics. International Organization for Standardization, Geneva.

ISO 21597-1 (2020). Information container for linked document delivery – Exchange specification – Part 1: Container. International Organization for Standardization, Geneva.

ISO 21597-2 (2020). Information container for linked document delivery – Exchange specification – Part 2: Link types. International Organization for Standardization, Geneva.

ISO 29481-1 (2016). Building information models – Information delivery manual – Part 1: Methodology and format. ISO/TC 59/SC 13 Organization and digitization of information about buildings and civil engineering works, including building information modelling (BIM). International Organization for Standardization, Geneva.

ISO 8000-1 (2022). Data quality – Part 1: Overview. International Organization for Standardization, Geneva.

ISO/IEC 19505 (2012). Information technology – Object Management Group Unified Modeling Language (OMG UML). International Organization for Standardization, Geneva.

ISO/IEC 19505-2 (2012). Information technology – Object Management Group Unified Modeling Language (OMG UML) – Part 2: Superstructure. International Organization for Standardization, Geneva.

ISO/TS 10303-1214 (2004). Industrial automation systems and integration – Product data representation and exchange – Part 1214: Application module: System breakdown. International Organization for Standardization, Geneva.

ISO/TS 10303-1216 (2008). Industrial automation systems and integration – Product data representation and exchange – Part 1216: Application module: Functional breakdown. International Organization for Standardization, Geneva.

Long, D. and Scott, Z. (2011). *A Primer for Model-Based System Engineering*. Vitech, Blacksburg.

Miller, J. and Mukerji, J. (2003). MDA Guide Version 1.0. OMG Document: omg/2003-05-01.

MOF (2014). Specification v2.4.2. OMG Document [Online]. Available at: https://www.omg.org/spec/MOF/2.4.2. [Accessed 14 February 2023].

OMG (2002). Systems Modelling Language (SysML), Version 1.2 [Online]. Available at: http://www.sysml.org/docs/specs/OMGSysML-v1.2-10-06-02.pdf [Accessed 14 February 2023].

Roxin, A. (2018). Reasoning and knowledge – Tradeoff between expressivity and efficiency. Habilitation to Conduct Research, Université Bourgogne Franche-Comté (UBFC), Dijon [Online]. Available at: https://hal.archives-ouvertes.fr/tel-01940781

Rumbaugh, J., Booch, G., Jacobson, I. (1999). *The Unified Modelling Language Reference Manual*. Addison-Wesley, Wokingham.

Steinberg, D., Budinsky, F., Paternostro, M. (2003). *Eclipse Modelling Framework*. Addison-Wesley, Wokingham.

Tuyen, F. (2006). *The Fast Guide to Model Driven Architecture. The Basics of Model Driven Architecture (MDA)*. Cephas Consulting Corp, Garden Valley.

XMI (2014). XML Metadata Interchange Version 1.1. Object Management Group, Needham [Online]. Available at: http://www.omg.org/spec/XML/ [Accessed 14 February 2023].

6

Building Information Modeling and Lean Construction

Lauri KOSKELA[1], Saeed TALEBI[2], Algan TEZEL[3] and Patricia TZORTZOPOULOS[1]

[1]*School of Arts and Humanities, University of Huddersfield, United Kingdom*

[2]*School of Engineering and Built Environment, Birmingham City University, United Kingdom*

[3]*College of Engineering & Physical Sciences, Aston University, Birmingham, United Kingdom*

6.1. Introduction

The recognition of the connection between building information modeling (BIM) and Lean Construction is relatively recent. Although these fields share a common objective, namely to improve the efficiency of construction, they had been advanced by their own communities of researchers and practitioners, with practically no interaction. The synergy between these fields was first spotted on the ground by a contractor with simultaneous initiatives in BIM and Lean (Markgraff 2008):

> Tocci Building Companies uses BIM as the cornerstone of its lean activities.
>
> We're finding that BIM is the foundational tool for implementing an efficient delivery process.

The idea of this synergy was surprising and intriguing and led to the first journal paper specifically addressing it in 2010 (Sacks et al. 2010a). In that article, 56 interactions between BIM and Lean were found. This seminal paper has now (February 2023) been cited around 900 times. Which provides evidence on the rapid proliferation of the topic. Especially, this synergy has been analyzed in more specific areas, such as operations and maintenance (Guzman and Ulloa 2020), mechanical, electrical and plumbing engineering (Tillmann 2020), demolition activities (Marzouk et al. 2019) and facilities management (Terreno et al. 2019).

The aim of this chapter is to give an overview and analysis on the connection between BIM and Lean. It is structured as follows: The next section endeavors to give a high-level overview on the component parts of the topic and their relations. The following sections discuss the contributions of BIM to Lean, respectively, in design, construction and facilities maintenance. The contributions of Lean to BIM are discussed in the subsequent section, and the chapter is completed by a conclusions section.

6.2. Overview on BIM and Lean

6.2.1. *Building information modeling*

The US National Building Information Model Standard Project Committee defines BIM as follows (National Institute of Building Sciences 2021):

> Building Information Modeling (BIM) is a digital representation of physical and functional characteristics of a facility. A BIM is a shared knowledge resource for information about a facility forming a reliable basis for decisions during its lifecycle; defined as existing from earliest conception to demolition.

According to Eastman et al. (2008, synopsis on the back cover), BIM is "a new approach to design, construction, and facilities management, in which a digital representation of the building process is used to facilitate the exchange and interoperability of information in digital format".

These definitions deserve to be expanded for some critical aspects. Building Information Models usually have an object-oriented, parametric data structure for the geometry of designed entities. Other information can be attached to such objects. The models are solid, that is, they are three-dimensional models.

Because of their characteristics, Building Information Models provide a central database and a multitude of functionalities that are new in comparison to the previous situation where information was presented on drawings and documents, leading to duplication of information across diverse documents, and great potential for errors.

6.2.2. *Lean*

Lean Construction can be understood in terms of tools, principles and theoretical foundations.

Tools provide the best known and practice-facing aspect of Lean, aiming to reduce waste in processes and generate better value to customers. A representative list of tools is provided by Umstot and Fauchier (2017):

- collaboration;
- 5S (a method for creating order on work place);
- Last Planner System (LPS);
- visual work place;
- root cause analysis;
- A3 (a problem-solving method);
- value stream mapping;
- Target Value Design;
- Just-in Time.

In turn, principles provide rules which have to be interpreted in the context and situation at hand; the methods and tools listed above usually embody one or more principles. Sacks et al. (2010a) discuss a number of principles of Lean Construction, summarized in Table 6.1.

Theories and foundations provide the explanation to Lean Construction. Koskela (2020) argued that Lean is a theoretical and philosophical innovation on three frontiers: (1) theory of production, (2) ontology (branch of philosophy addressing what is there in the world) and (3) epistemology (branch of philosophy addressing how knowledge is acquired). These are explained in the following.

In terms of the theory of production, it is claimed (Koskela 2020) that Lean means a shift from the transformation theory of production to two other theories, namely the flow theory and the value generation theory. The transformation theory looks at production as a black box; the flow theory brings time and uncertainty, both being causes of waste, into the analysis of production, while the value generation theory adds the customer and value into the picture.

Focus area	Principles
Flow	Reduce variability Reduce cycle times Reduce batch sizes (strive for single piece flow) Increase flexibility Select an appropriate production control approach Standardize Institute continuous improvement Use visual management Design the production system for flow and value
Value generation process	Ensure comprehensive requirements capture Focus on concept selection Ensure requirement flowdown Verify and validate
Problem-solving	Go and see for yourself Decide by consensus, consider all options
Developing partners	Cultivate an extended network of partners

Table 6.1. *Principles of Lean Construction according to Sacks et al. (2010a)*

The philosophical foundations of Lean embrace ontological and epistemological considerations (Koskela 2020). Regarding ontology, Lean is supported by process ontology (Koskela and Kagioglou 2005), which stresses the changing as well as the relational nature of phenomena. This replaces the mainstream approach of thing ontology, based on the idea of stability and decomposability of things. Regarding epistemology, Lean is compatible with Aristotelianism (Koskela et al. 2019), addressing both the derivation of knowledge from the empirical world, and the deductive use of existing knowledge. The mainstream counterpart is Platonism, which emphasizes the one-directional deductive use of existing knowledge.

6.2.3. *Relation between BIM and Lean*

The connections between Lean and BIM are multiple and of various characters (Sacks et al. 2010a). It is thus opportune to present a simplified overview on them. This is possible by looking at the theoretical and philosophical foundations of Lean, and the compatibility of the functionalities of BIM with them.

Before proceeding further, it is important to stress that the interaction between Lean and BIM is claimed to be mutual: BIM supports Lean, and Lean supports BIM (Figure 6.1). Originally, the focus was almost exclusively on how BIM supports Lean, and the other direction was overlooked.

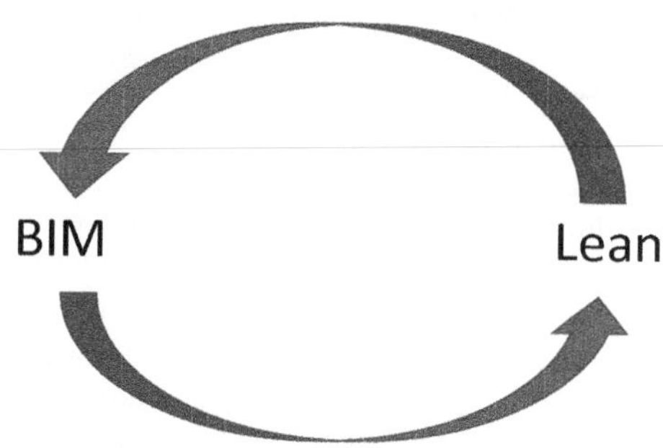

Figure 6.1. *The relations between BIM and Lean*

6.2.3.1. *BIM supporting Lean*

Now, the rapid proliferation of BIM can be associated to its extraordinary compatibility with the theoretical and philosophical foundations of Lean, outlined in Table 6.2. BIM is instrumental in removing problems causing uncertainty and wasteful use of time. It facilitates client evaluation of design solutions through better visualization and understanding. Furthermore, BIM supports process ontology by facilitating collaboration to tackle the relational nature of design and production, and by allowing continuous improvement. Finally, BIM allows simulated experimentation, in the spirit of Aristotelian epistemology. In all these aspects, the previous counterpart of BIM, in terms of paper drawings and documentation, was remarkably weak.

6.2.3.2. *Lean supporting BIM*

It is contended that when BIM is implemented into a design and construction context where Lean principles have also been applied, the results will be better than otherwise. The Lean Construction context arguably facilitates the implementation of BIM, especially in its initial stages, and generally supports the achievement of correct and useful models. This is especially the case with firms adopting Lean features such as discipline and predictability, collaboration and experimentation, as well as continuous improvement. Moreover, Lean principles have turned out to be applicable in information systems work in general, and related practices are usable also in regard to BIM.

<table>
<tr><th>Area of theory/philosophy</th><th>Specific theory/philosophical approach</th><th>Contributions of BIM (and associated technologies)</th></tr>
<tr><td rowspan="2">Theory of production</td><td>Flow theory</td><td>Removing problems (e.g. through clash detection)</td></tr>
<tr><td>Value generation theory</td><td>Facilitating client evaluation of designs (e.g. through visualization)</td></tr>
<tr><td>Ontology</td><td>Process ontology</td><td>Supporting collaboration (e.g. through easier sharing of information and by providing common ground)
Supporting continuous improvement (e.g. through rapid generation of alternative solutions)</td></tr>
<tr><td>Epistemology</td><td>Aristotelian epistemology</td><td>Supporting evaluation of solutions (especially through analysis and simulation)</td></tr>
</table>

Table 6.2. *Theoretical and philosophical foundations of Lean and contributions of BIM related to them*

6.3. Contributions of BIM to Lean in design, construction and facilities maintenance

6.3.1. *BIM for Lean in design*

The synergies between BIM and Lean are perhaps most apparent during the design phase. In early design, where the scope and feasibility of a project are determined, BIM enables rapid creation, communication and comparison of multiple design alternatives (Sacks et al. 2010a). Supported by BIM's advanced visualization and simulation capabilities, this facilitates the practical execution of Set-based Design, a Lean design management approach that advocates keeping design options flexible for as long as possible during the development process instead of choosing a single point solution upfront (Lee et al. 2012). This is argued to support verification and validation, and focusing on concept selection for project value (Parrish 2009; Tzortzopoulos et al. 2020). Some necessary non-value adding activities in the early design phase such as creating bill of quantities (BOQ) for different design alternatives are automated by BIM tools, which reduces the risks associated with mistakes in estimating and overall procurement durations (Khosakitchalert et al. 2020). When model elements are linked with accurate cost data, automated cost estimates by different design alternatives can be rapidly generated (Smith 2016), which guides procurement decisions and facilitates the execution of the lean design management practice of Target Value Design, a collaborative design development process focusing on delivering project value for or below a targeted project cost (Ballard 2008).

Alongside the rapid generation of alternatives and BOQs, the automated production of design drawings, documents and coordinated views on BIM models helps reduce design cycle-times (Eastman et al. 2011). Those BIM features prevent duplicated efforts, reduce risks associated with design documentation (e.g. mistakes in drawings) and automate some necessary non-value adding tasks such as the need for preparing different sets of design drawings, for example, plans, facades and controlling consistency across those drawing sets/views (Eastman et al. 2011). This enables designers to focus their efforts on generating design solutions for greater project value. The ability to automatically generate and navigate the model views at will supports information flows and information-pull (on demand information) for project stakeholders.

BIM also streamlines the design process with fewer mistakes performed through (Vermeulen and Ayoubi 2019) (i) parametric design (i.e. the user defines associations between design elements and geometries – "this window depends upon this wall and will move with it"), (ii) design automation (i.e. the ability to automate tasks within parametric models with scripts created on, for instance, Autodesk's Dynamo – "create a door for every x meter of a wall") and (iii) computational modeling features (i.e. the user explicitly describes a process to create a design outcome – "create a number of windows on the facade and evaluate how many exits per unit area are needed").

Design technology is progressing toward the generative design model, where describing a set of design goals and constraints (e.g. minimum amount of light in a particular space, maximum amount to be spent for construction) leads to automated generation and optimization of multiple-design options (see for instance Autodesk's Project Refinery). Supported by other technologies such as cloud computing and artificial intelligence (AI), parametric design and design automation capabilities of BIM are enabling this progression, which will further help realize Lean ideals, that is, better concept selection, verification and requirements capture (Sydora and Stroulia 2020). Another fast-progressing area in design automation leading to less necessary non-value adding efforts for designers and improved design quality with better standardization is automatic rule-based code checking and compliance. This involves assessing design compliance to codes and regulations in automated or semi-automated ways (Bloch and Sacks 2018; Sydora and Stroulia 2020).

The collaborative design processes enabled by BIM help minimize delays, mistakes and misunderstandings, to improve information flow and design process control, reduce variability and compress the overall design cycle time (Dave et al. 2013). The current practice in the industry is for different design disciplines (i.e. architectural, structural, MEP) to create their own models, which have to be periodically combined and validated (federated/coordinated). Although useful, this gives rise to issues such as model interoperability, need for specific data creation and exchange protocols, and considerations for intellectual property rights (Dave et al. 2013).

In line with collaborative design processes, project teams can jointly hold iterative modeling coordination meetings to carry out design reviews, to monitor design status (e.g. checked, approved, final), control and visualize design revisions, and address clashes, errors and omissions that have been identified in the design (Eastman et al. 2011; Tauriainen et al. 2016). This supports information flows, reduction of variability and increased standardization in the design phase. Those iterative meetings enabled by BIM are also necessary for Target Value Design (Pishdad-Bozorgi et al. 2013; Do et al. 2014; Tauriainen et al. 2016).

Geometric and semantic-/rule-based clash detection and management of hard and soft clashes have been widely described as a "low hanging fruit" benefit from BIM, which is relatively easy to achieve in practice. Addressing the clashes early in the design phase increases design quality, compresses project delivery times and reduces re-work in the construction phase (Tauriainen et al. 2016).

Early stakeholder involvement increases BIM benefits toward greater project value and better information flow. BIM information can be used downstream by contractors, commercial managers/quantity surveyors and suppliers in processes such as estimating, detailing for fabrication/off-site construction, site planning, production planning (4D, i.e. 3D models linked with construction schedule) and resource planning (5D, i.e. 4D linked to costs and resources) (Dave et al. 2013). BIM streamlines workflows for stakeholders as well. For instance, the machine-readable nature of BIM models facilitates the automatic generation of detailed drawings, fabrication of complex designs on computer numerical control (CNC) machines and coordination of logistics for prefabricated elements (Hardin and McCool 2015).

Immersive technologies such as virtual reality (VR)/mixed reality (MR) are now more frequently used with BIM in client and stakeholder engagement for the communication of design intent, requirement capture as well as coordination and verification through advanced visualization (Zaker and Coloma 2018; Wang et al. 2018; Ergün et al. 2019).

BIM allows the design model to be simulated against performance criteria. Simulation and analysis of the models through different performance parameters such as architectural/spatial layout, seismic, acoustic, thermal, ventilation, energy/sustainability, lifecycle cost, constructability indicators can be automatically executed by BIM tools

(Dave et al. 2013). Alongside many positive synergies, increased complexity in the management of both Lean and BIM processes, increased inventory of alternative designs and design drawings can be listed as the potential negative synergies between BIM and Lean in the design phase (Sacks et al. 2010a).

6.3.2. *BIM for Lean Construction*

The use of BIM for Lean Construction efforts has been widening in the construction (production) phase of the project lifecycle. This manifests in two forms: (i) BIM helps realize some Lean principles and (ii) BIM supports and enhances some Lean Construction techniques. One of the prominent Lean techniques where the interaction with BIM is apparent is the LPS for construction planning and control. The LPS is a collaborative project planning process that involves trade foremen or design team leaders (the last planners) in the planning and control process in greater detail as the time for the work to be done gets closer from master plans to lookahead and weekly plans (Ballard 2000). It supports construction production planning and control by providing systematic routines to increase workflow reliability and process stability. In the United Kingdom, it is known as Collaborative Planning, and in the United States, it is sometimes called Pull Planning (Daniel et al. 2017).

In the LPS, BIM is used for 4D planning during the master scheduling (long-term) to select, sequence and size work, and for 3D visualization of construction processes, location-based planning, constructability analysis and clash detection during the lookahead (medium-term) and weekly (short-term) work planning to make work ready by screening and pulling (Bhatla and Leite 2012; Garrido et al. 2015). Online communication of product and process information through BIM also supports continuous improvement efforts within the LPS, where site teams take note of the discrepancies between designs, plans and actual site conditions on mobile devices (Tillman and Sargent 2016). Commercial software facilitating this LPS integration with BIM workflows, such as VisiLean (Dave et al. 2011), are becoming available. There are also prototypes that enable enhanced functionalities through the combined use of BIM and the LPS such as Lean Enterprise Web-based Information System (LEWIS) (Spripraset and Dawood 2003), the "pull" based and visual construction work planning and control system KanBIM (Sacks et al. 2013), Smart Construction Planner (Guerriero et al. 2017) and BIM-based Last Planner System (Heigermoser et al. 2019).

At the Lean principles level, the BIM integration with the LPS contributes to (Hamdi and Leite 2012) (i) model-based coordination across different construction disciplines, which increases planning reliability, reduces planning cycles times and supports shared understanding of work for better constraint identification; (ii) reduction of variability, waste, errors and conflicts in site operations and (iii) verification and validation of design and construction processes for the first site operations.

Alongside 4D modeling and clash detection, the BIM functionality of model-based, automatic quantity take-offs reduces the amount of non-value adding activities (waste) and errors in work planning and progress calculations (Bryde et al. 2013; Monteiro and Poças Martins 2013; Hardin and McCool 2015). Furthermore, the programmable nature of BIM models gives way to many rule-based automation and simulation opportunities for construction such as automatic site layout planning (Kumar and Cheng 2015; Schwabe et al. 2019), site safety (Zhang et al. 2013) and hazard identification (Zhang et al. 2015), real-time construction quality (Wang et al. 2015) and progress control (Han and Golparvar-Fard 2014; Golparvar-Fard et al. 2015), automatic production of documents for quality management (Chen and Luo 2014), monitoring and tracking of logistics operations (Irizarry et al. 2013), on-site monitoring of carbon emissions with the use of BIM systems (Wong and Zhou 2015; Chong et al. 2017), and BIM-based demolition and renovation waste estimation and planning (Cheng and Ma 2013), saving the time spent in those unproductive tasks while reducing the subjectivity and variability associated with them.

The visualization capacity of BIM offers new avenues for maintaining the information flow and creating transparency in construction processes across different disciplines and organizational levels. Gerber et al. (2010) demonstrate examples from case studies around the world such as communicating the complex erection sequences of a solar chimney with on-site teams through detailed snapshot images from the chimney's BIM-based work package or detailed BIM outputs on precast elements, enabling on-site teams to more easily and precisely "pull" what they need from information provided by engineering teams.

In large scale infrastructure projects, BIM models and their outputs are used to support the daily huddle meetings with site teams, and to maintain the coordination between subcontractors working dispersedly in different areas (Tezel and Aziz 2017). Virtual *Gemba* walks – management's observation of the site conditions and site walkthroughs for work coordination (Mahalingam et al. 2015), creating a shared understanding (Johansson et al. 2015), and quality and safety analyses (Zou et al. 2017) with project teams are realized on BIM models. BIM also constitutes the technological foundation for the *Obeya* (Big Room) concept, where co-located project teams (integrated through relational contracts) manage and execute construction projects through direct, face-to-face interaction over BIM models, reducing delays in decision-making and problems in communication (Dave et al. 2015). On-site communication platforms, together with mobile devices integrated with cloud-based BIM viewers, BIM-stations (Vestermo et al. 2016) or BIM-kiosks (Bråthen and Moum 2016), are employed for the teams in the field to pull design information from the BIM models when they need it.

The growing demand for engineer-to-order (ETO) prefabricated systems in the construction industry aligns well with the Lean principles and ideals. From a broader perspective to off-site construction, BIM helps to establish better connections between off-site manufacturers and construction sites by allowing construction (and design) data to be machine processable and components to be manufactured without human intervention (Eastman and Sacks 2008). This results in greater precision in specifying material requirements and dimensions, which can reduce over-ordering and thus decrease construction site waste, particularly regarding complex structures (Abanda et al. 2017). In addition, BIM can assist manufacturers and contractors by providing a 3D model of off-site element positions, connections and construction (Abanda et al. 2017).

The ability to integrate BIM data with advanced hardware such as 3D laser scanners to collect as-built data and establish a point cloud model for system coordination supports site operations with robotic total stations (Zhang et al. 2016). In this domain, BIM also helps in resolving complexities in logistics planning and control for site assembly of ETO prefabricated systems through 4D modeling (Bortolini et al. 2019). The wealth of information contained within or linked to BIM models enables the possibility for bridging the interfaces and continuous information flow between designers, suppliers, manufacturers and users for prefabricated systems (Ezcan et al. 2013).

6.3.3. *BIM for Lean facilities management*

A facility can be defined as "a collection of assets built, installed or established to serve the needs of an entity (people or an organization)" (International Standards Organization 2017, p. 3). Examples of assets can be infrastructure, real estate and utilities. Facility management (FM) is defined as "the effective management of place and space, integrating an organization's support infrastructure to deliver services to staff and customers at best value whilst enhancing overall organizational performance" (Royal Institute of Chartered Surveyors 2020, p. 7). FM consists of several multidisciplinary activities and is essential to achieve success in any project and organization (Noor and Pitt 2009)

Information management is recognized as the crux of effective FM and consists of delivering accurate, timely and relevant information (Terreno et al. 2019). In the conventional practice of FM, a significant amount of time is wasted and a great deal of extra work is carried out to search for the required and relevant information. Moreover, the generated information may not be used because the information was not created in time, relevant information may not be found among irrelevant information or relevant information is ignored (Jylhä and Suvanto 2015). The use of BIM in FM can enable consistent and coordinated information exchange over a facility's lifecycle, from design to maintenance and operation.

In particular, BIM with its visualization and analysis capabilities is conceived as an impetus to provide accurate information to FM systems in time and enhance other functions of FM (Becerik-Gerber et al. 2012). Such capabilities in BIM can aid FM by enabling localization of facility components and indoor navigation (Volk et al. 2014). However, the adoption of BIM for FM has been slow compared to the adoption of BIM in design and construction (Talebi 2014a, 2014b). According to Volk et al. (2014), one of the major reasons behind such slow adoption is the difficulty in updating and maintaining information in BIM during the FM stage as stakeholders are not often willing to pay for this purpose. This challenge is originally traced to the deficient development of theory in FM (Edirisinghe et al. 2017).

The Transformation, Flow and Value (TFV) theory of production (Koskela 2000) has been used to explain the difficulties in implementing BIM in FM (Shou et al. 2014; Munir et al. 2019). From the TFV theory perspective, even though the focus of FM has shifted from cost minimization in real estate operations to supporting end-customer requirements using

BIM, the existing literature on BIM-enabled FM does not explain the customer value generation process or information flows between stakeholders (Munir et al. 2019). In other words, according to the existing literature, BIM-enabled FM is mainly focused on the "T" view (Jylhä and Junnila 2013). BIM is changing the operation of built assets (Love et al. 2014) by focusing on pointwise improvements (e.g. data visibility, data analytics, maintenance audit, service procurement) (Munir et al. 2019). Arguably, such a focus on "T" view is the cause of waste and value loss and does not provide a basis for managing the value creation challenges (Jylhä and Junnila 2013) or improving the flow of information (i.e. updating BIM during FM) (Lee and Akin 2009). A review of the literature reveals that some of the principles from Lean Construction are already adopted in FM, but they have not yet been widely implemented in this field to balance the transformation, flow and value views. Streamlining processes is proven to be effective toward adopting Lean principles and facilitating the implementation of BIM in FM (Kasprzak and Dubler 2012).

Problems of interoperability between BIM and FM (Love et al. 2015) and variability in information management (Bascoul et al. 2018) are significant challenges and result in unnecessary, excessive, irrelevant or defective information (Jylhä and Suvanto 2015). BIMs developed during design and construction do not contain a significant amount of required information for FM and also such information does not necessarily contribute to FM (Bonanomi 2016a) due to the lack of clear requirements for the adoption of BIM in FM (Edirisinghe et al. 2017). In other words, the utilization of BIM from design to maintenance and operation is deficient (Kiviniemi and Codinhoto 2014).

Standardization is proposed to resolve the barriers with interoperability and variability in information management. A number of standards, such as Industry Foundation Class (IFC) and Construction Operation Building Information Exchange (COBie), have been developed to support the concepts of interoperability and integration (Pärn et al. 2017). Moreover, Succar and Poirier (2020) suggest to use the concept of model-based deliverables for standardizing and clarifying the required information for the adoption of BIM in FM.

A primary technique of the Lean Construction is "pull" (Ballard 1999). Sacks et al. (2010b) discuss the potential contribution of BIM to enable the "pull" technique to reduce variability in information. The pull technique (Womack et al. 1990) in the context of FM means that only such information should be produced upstream that the customer downstream needs to operate and maintain the facility (Succar 2009). The pull mechanism results in adequate, accurate and timely information (Ghosh et al. 2015). The existing practice of BIM-enabled FM still heavily relies on the push mechanism, which depends on forecasts to determine what information, when and how much should be generated (Becerik-Gerber et al. 2012). Ideally, the facility manager should enter the design stage early, influence the design and construction and help produce a BIM that pulls the requirements of the customer (Azhar 2011), including those related to FM. In other words, early involvement of the facility manager leads to an understanding of what information should be modeled (Bonanomi 2016b) with the major focus on the operation and maintenance (Kasprzak and Dubler 2012).

Continuous improvement (CI), which is at the heart of Lean (Womack et al. 1990), is essential to the success of any FM organization (Beck et al. 2016). CI is an approach by which (a) small incremental improvement steps are taken to improve performance (Slack et al. 2010) and (b) waste in all processes of an organization are identified, reduced and eliminated (Bessant et al. 2001). Standardization and the pull technique are the foundation of continuous improvement (Gao et al. 2020). In particular, continuous improvement is essential in service provision due to the rising demand for quality FM services from customers (Smith 2010).

In conclusion, BIM is known as a solution to provide consistent and coordinated information exchange between stakeholders during the facility's lifecycle. However, the adoption of BIM for facilities management has been slow. Based on the TFV theory of production, it has been explained by the existing literature that BIM-enabled FM has been focused on the "T" view, which does not cover the customer value generation process or the flow of the information between stakeholders. It was further discussed that reduction of variability, standardization and continuous improvement, which fall under the "F" category, as well as comprehensive requirements capture and requirements flowdown, which fall under the "V" category, seem essential to facilitate the implementation of BIM for Lean operation and facilities management.

6.4. Lean for BIM

Up to now, the focus has been on the support that BIM can provide for the implementation of Lean. Here, the reverse relation is also considered:

how can Lean support the implementation of BIM? This topic embraces two viewpoints: (1) how Lean can be applied in information system work (into which BIM partially falls) generally, and (2) how the Lean design and construction context supports the implementation and continuous improvement of BIM. The former viewpoint is related to the hardware and software viewpoints to BIM, whereas the latter deals with the design and construction processes into which BIM is embedded.

The use of Lean in IT work has attracted attention in the last 10 years, and practice-based guidelines have emerged (Plenert 2011; Bell and Orzen 2016; Williams and Duray 2017). These sources stress that Lean principles and tools apply to work on IT systems similarly to other contexts. At the most general level, the importance of value to the customer and waste elimination accentuate. Complete processes (where IT support is developed or maintained) should be addressed. The A3 problem-solving method is recommended as a central approach in the design of information systems (for a thorough introduction of the A3 method, see Sobek II and Smalley 2011). In turn, the plan-do-check-act (PDCA) cycle offers itself for evaluating designed solutions and for continuous improvement. Work on existing information systems tends to be reactive – solving emerging problems in the functioning of the systems. The Lean approach implies a proactive approach for realizing the target state.

The argument is that when BIM is implemented into a design and construction context where Lean principles have also been applied, the results will be better than otherwise; for initial evidence see (Mahalingam et al. 2015). Arguably, there is an analogy to computer-integrated manufacturing (CIM): "CIM acts as a magnifying glass. It makes the good system much better; it makes the poor system much worse" (Melnyk and Narasimhan 1992, p. 91). Such characteristics of the Lean Construction context as discipline, collaboration, experimentation, as well as continuous improvement are arguably paramount for facilitating both the initial and mature implementations of BIM. Discipline and collaboration are needed for the gradual build-up of correct and useful models; for this, the LPS of production control has turned out to be an effective tool (Bhatla and Leite 2012; Mäki and Kerosuo 2020). Experimentation, realized through short cycles of PDCA, allows for exploring and trialing new possibilities of using the functionalities of a BIM. In turn, continuous improvement, likewise supported by the PDCA cycle, helps solving the various problems unavoidably emerging when new technology is implemented, and also facilitates the exchange of lessons learnt among projects.

6.5. Conclusion

As discussed in the previous sections, the large number of connections between BIM and Lean, which have been discovered in the last 10 years, leads to one conclusion: construction projects, as well as companies involved in them, should simultaneously and synergistically implement both Lean and BIM. The situation is dynamic and evolutionary: the capabilities of BIM are constantly developing, and both understanding and practice of Lean are improving. This means that the possibilities for synergistic interaction between Lean and BIM are also deepening.

In view of this situation, the interaction between BIM and Lean continues to be a fertile topic for research. Up to now, the interest has mostly addressed how BIM can support Lean principles, and in explorative research, a multitude of related mechanisms have been found. It is now also instrumental to launch exploitative activities for codifying such existing knowledge into practical guidelines. Regarding how Lean principles can support BIM, more explorative research is still needed.

6.6. References

Abanda, F.H., Tah, J.H.M., Cheung, F.K.T. (2017). BIM in off-site manufacturing for buildings. *Journal of Building Engineering*, 14, 89–102.

Azhar, S. (2011). Building information modeling (BIM): Trends, benefits, risks, and challenges for the AEC industry. *Leadership and Management in Engineering*, 11(3), 241–252.

Ballard, G. (1999). Can pull techniques be used in design management? CIB Report, Concurrent Engineering in Construction, Helsinki.

Ballard, G. (2000). The last planner system of production control. Thesis, University of Birmingham, Birmingham.

Ballard, G. (2008). The lean project delivery system: An update. *Lean Construction Journal*, 1–19.

Bascoul, A.M., Tommelein, I.D., Tillmann, P., Muxen, S. (2018). Towards facility management participation in design: A UCSF case study. In *26th Annual Conference of the International Group for Lean Construction*, Chennai.

Becerik-Gerber, B., Jazizadeh, F., Li, N., Calis, G. (2012). Application areas and data requirements for BIM-enabled facilities management. *Journal of Construction Engineering and Management*, 138, 431–442.

Beck, S., Schmalz, S., Heyl, J.V., Binder, F. (2016). Optimizing the value stream-application of BIM in FM. Status quo in Germany. In *24th Annual Conference of the International Group for Lean Construction*, Boston, MA.

Bell, S.C. and Orzen, M.A. (2016). *Lean IT: Enabling and Sustaining your Lean Transformation*. CRC Press, New York.

Bessant, J., Caffyn, S., Gallagher, M. (2001). An evolutionary model of continuous improvement behaviour. *Technovation*, 21, 67–77.

Bhatla, A. and Leite, F. (2012). Integration framework of BIM with the last planner system TM. In *IGLC 20th Conference of the International Group for Lean Construction*, San Diego, CA.

Bloch, T. and Sacks, R. (2018). Comparing machine learning and rule-based inferencing for semantic enrichment of BIM models. *Automation in Construction*, 91, 256–272.

Bonanomi, M. (2016a). Building information modeling (BIM) and facility management (FM). In *Knowledge Management and Information Tools for Building Maintenance and Facility Management*, Talamo, C. and Bonanomi, M. (eds). Springer, Milan.

Bonanomi, M. (2016b). Methodological experimentation: Proposal of a datasheet template for FM activities in the BIM environment. In *Knowledge Management and Information Tools for Building Maintenance and Facility Management*, Talamo, C. and Bonanomi, M. (eds). Springer, Milan.

Bortolini, R., Formoso, C.T., Viana, D.D. (2019). Site logistics planning and control for engineer-to-order prefabricated building systems using BIM 4D modeling. *Automation in Construction*, 98, 248–264.

Bråthen, K. and Moum, A. (2016). Bridging the gap: Bringing BIM to construction workers. *Engineering, Construction and Architectural Management*, 23, 751–764.

Bryde, D., Broquetas, M., Volm, J.M. (2013). The project benefits of Building Information Modelling (BIM). *International Journal of Project Management*, 31, 971–980.

Chen, L. and Luo, H. (2014). A BIM-based construction quality management model and its applications. *Automation in Construction*, 46, 64–73.

Cheng, J.C.P. and Ma, L.Y.H. (2013). A BIM-based system for demolition and renovation waste estimation and planning. *Waste Management*, 33, 1539–1551.

Chong, H.-Y., Lee, C.-Y., Wang, X. (2017). A mixed review of the adoption of Building Information Modelling (BIM) for sustainability. *Journal of Cleaner Production*, 142, 4114–4126.

Daniel, E.I., Pasquire, C., Dickens, G., Ballard, H.G. (2017). The relationship between the Last Planner® System and collaborative planning practice in UK construction. *Engineering, Construction and Architectural Management*, 24, 407–425.

Dave, B., Boddy, S., Koskela, L. (2011). Visilean: Designing a production management system with lean and BIM. In *19th Annual Conference of the International Group for Lean Construction*, Lima.

Dave, B., Koskela, L., Kiviniemi, A., Owen, R.L., Tzortzopoulos Fazenda, P. (2013). Implementing Lean in construction: Lean construction and BIM. Guide C725, CIRIA, London.

Dave, B., Pikas, E., Kerosuo, H., Mäki, T. (2015). ViBR – Conceptualising a virtual big room through the framework of people, processes and technology. *Procedia Economics and Finance*, 21, 586–593.

Do, D., Chen, C., Ballard, G., Tommelein, I. (2014). Target value design as a method for controlling project cost overruns. *International Group for Lean Construction*, 22, 171–181.

Eastman, C.M. and Sacks, R. (2008). Relative productivity in the AEC industries in the United States for on-site and off-site activities. *Journal of Construction Engineering and Management*, 134, 517–526.

Eastman, C.M., Teicholz, P., Sacks, R., Liston, K. (2008). *BIM Handbook: A Guide to Building Information Modeling for Owners, Managers, Architects, Engineers, Contractors, and Fabricators*. John Wiley & Sons, Hoboken, NJ.

Eastman, C.M., Teicholz, P., Sacks, R., Liston, K. (2011). *BIM Handbook: A Guide to Building Information Modeling for Owners, Managers, Designers, Engineers and Contractors*. John Wiley & Sons, Hoboken, NJ.

Edirisinghe, R., London, K.A., Kalutara, P., Aranda-Mena, G. (2017). Building information modelling for facility management: Are we there yet? *Engineering, Construction and Architectural Management*, 24(6), 1119–1154. [Online]. Available at: https://doi.org/10.1108/ECAM-06-2016-0139.

Ergün, O., Akın, Ş., Dino, İ.G., Surer, E. (2019). Architectural design in virtual reality and mixed reality environments: A comparative analysis. *IEEE Conference on Virtual Reality and 3D User Interfaces (VR)*, Osaka, 914–915, doi: 10.1109/VR.2019.8798180.

Ezcan, V., Isikdag, U., Goulding, J. (2013). BIM and off-site manufacturing: Recent research and opportunities. CIB World Building Congress 2013. TG 74 Workshop.

Gao, S., Pheng, L.S., Tay, W. (2020). Lean facilities management: Preliminary findings from Singapore's international schools. *Facilities*, 38, 539–558.

Garrido, M., Mendes, R., Scheer, S., Campestrini, T. (2015). Using BIM for last planner system: Case studies in Brazil. In *Proceedings of the 2015 International Workshop on Computing in Civil Engineering*, Austin, TX.

Gerber, D.J., Becerik-Gerber, B., Kunz, A. (2010). Building information modeling and lean construction: Technology, methodology and advances from practice. In *Proceedings of the 18th International Group for Lean Const*, Haifa, 14–16.

Ghosh, A., Chasey, A.D., Mergenschroer, M. (2015). Building information modeling for facilities management: Current practices and future prospects. In *Building Information Modeling: Applications and Practices*. American Society of Civil Engineers, Reston, VA.

Golparvar-Fard, M., Pena-Mora, F., Savaresse, S. (2015). Automated progress monitoring using unordered daily construction photographs and IFC-based building information models. *Journal of Computing in Civil Engineering*, 29, 04014025.

Guerriero, A., Kubicki, S., Berroir, F., Lemaire, C. (2017). BIM-enhanced collaborative smart technologies for LEAN construction processes. In *International Conference on Engineering, Technology and Innovation (ICE/ITMC)*. IEEE, Madeira.

Guzman, G. and Ulloa, W. (2020). BIM application in the operation and maintenance management of a sports infrastructure. In *28th Annual Conference of the International Group for Lean Construction*, Tommelein, I.D. and Daniel, E. (eds). IGLC, Berkeley, CA.

Hamdi, O. and Leite, F. (2012). BIM and Lean interactions from the BIM capability maturity model perspective: A case study. In *Proceedings of the 20th Annual Conference of the International Group for Lean Construction*, San Diego, CA, 18–20.

Han, K.K., Golparvar-Fard, M. (2014). Automated monitoring of operation-level construction progress using 4D bim and daily site photologs. In *Proceedings of the Construction Research Congress: Construction in a Global Network (CRC 2014)*, Atlanta, GA, 19–21.

Hardin, B. and McCool, D. (2015). *BIM and Construction Management: Proven Tools, Methods, and Workflows*. John Wiley & Sons, New York.

Heigermoser, D., De Soto, B.G., Abbot, E.L.S., Chua, D.K.H. (2019). BIM-based Last Planner System tool for improving construction project management. *Automation in Construction*, 104, 246–254.

International Standards Organization (2017). Facility management-scope. Key concepts and benefits. ISO, Geneva.

Irizarry, J., Karan, E.P., Jalaei, F. (2013). Integrating BIM and GIS to improve the visual monitoring of construction supply chain management. *Automation in Construction*, 31, 241–254.

Johansson, M., Roupé, M., Bosch-Sijtsema, P. (2015). Real-time visualization of building information models (BIM). *Automation in Construction*, 54, 69–82.

Jylhä, T. and Junnila, S. (2013). Learning from lean management-going beyond input-output thinking. *Facilities*, 31, 454–467.

Jylhä, T. and Suvanto, M.E. (2015). Impacts of poor quality of information in the facility management field. *Facilities*, 33(5/6), 302–319.

Kasprzak, C. and Dubler, C. (2012). Aligning BIM with FM: Streamlining the process for future projects. *Construction Economics and Building*, 12, 68–77.

Khosakitchalert, C., Yabuki, N., Fukuda, T. (2020). Automated modification of compound elements for accurate BIM-based quantity takeoff. *Automation in Construction*, 113, 103142.

Kiviniemi, A. and Codinhoto, R. (2014). Challenges in the implementation of BIM for FM – Case Manchester Town Hall complex. In *Computing in Civil and Building Engineering*, 665–672.

Koskela, L. (2000). An exploration towards a production theory and its application to construction. Thesis, VTT Technical Research Centre of Finland, Espoo.

Koskela, L. (2020). Theory of lean construction. In *Lean Construction: Core Concepts and New Frontiers*, Tzortzopoulos, P., Kagioglou, M., Koskela, L. (eds). Routledge, London.

Koskela, L. and Kagioglou, M. (2005). On the metaphysics of production. In *13th Annual Conference of the International Group for Lean Construction*, Sydney.

Koskela, L., Ferrantelli, A., Niranen, J., Pikas, E., Dave, B. (2019). Epistemological explanation of lean construction. *Journal of Construction Engineering and Management*, 145, 04018131.

Kumar, S.S. and Cheng, J.C.P. (2015). A BIM-based automated site layout planning framework for congested construction sites. *Automation in Construction*, 59, 24–37.

Lee, S. and Akin, Ö. (2009). Shadowing tradespeople: Inefficiency in maintenance fieldwork. *Automation in Construction*, 18, 536–546.

Lee, S.-I., Bae, J.-S., Cho, Y.S. (2012). Efficiency analysis of set-based design with structural building information modeling (S-BIM) on high-rise building structures. *Automation in Construction*, 23, 20–32.

Love, P.E., Matthews, J., Simpson, I., Hill, A., Olatunjl, O.A. (2014). A benefits realization management building information modeling framework for asset owners. *Automation in Construction*, 37, 1–10.

Love, P.E., Matthews, J., Lockley, S. (2015). BIM for built asset management. *Built Environment Project and Asset Management*, 5(3), 230–232.

Mahalingam, A., Yadav, A.K., Varaprasad, J. (2015). Investigating the role of lean practices in enabling BIM adoption: Evidence from two Indian cases. *Journal of Construction Engineering and Management*, 141, 05015006.

Mäki, T. and Kerosuo, H. (2020). Design-related questions in the construction phase: The effect of using the Last Planner System in design management. *Canadian Journal of Civil Engineering*, 47(2), 132–139.

Markgraff, P. (2008). Leaning toward BIM? Green construction purchasing [Online]. Available at: http://www.construction-purchasing.com/site/archives/working_right/Markgraff_Q308_BIM.php.

Marzouk, M., Elmaraghy, A., Voordijk, H. (2019). Lean deconstruction approach for buildings demolition processes using BIM. *Lean Construction Journal*, 147–173.

Melnyk, S.A. and Narasimhan, R. (1992). *Computer Integrated Manufacturing.* Business One Irwin, Homewood, IL.

Monteiro, A. and Poças Martins, J. (2013). A survey on modeling guidelines for quantity takeoff-oriented BIM-based design. *Automation in Construction*, 35, 238–253.

Munir, M., Kiviniemi, A., Jones, S., Finnegan, S. (2019). BIM business value generation theory: A grounded theory approach. *Journal of Information Technology in Construction*, 24, 406–423.

National Institute of Building Sciences (2021). Frequently asked questions about the national BIM Standard – United States™. National BIM Standard – United States TM V3 [Online]. Available at: https://www.nationalbimstandard.org/faqs [Accessed 12/02/2023].

Noor, M. and Pitt, M. (2009), A critical review on innovation in facilities management service delivery. *Facilities*, 27(5/6), 211–228.

Pärn, E.A., Edwards, D.J., Sing, M.C. (2017). The building information modelling trajectory in facilities management: A review. *Automation in Construction*, 75, 45–55.

Parrish, K.D. (2009). Applying a set-based design approach to reinforcing steel design. Dissertation, University of California, Berkeley, CA.

Pishdad-Bozorgi, P., Moghaddam, E.H., Karasulu, Y. (2013). Advancing target price and target value design process in IPD using BIM and risk-sharing approaches. In *Proceedings of the 49th ASC Annual International Conference Proceedings*, San Luis Obispo, CA.

Plenert, G.J. (2011). *Lean Management Principles for Information Technology*. CRC Press, Boca Raton, FL.

Royal Institute of Chartered Surveyors (2020). Facilities management [Online]. Available at: https://www.rics.org/north-america/join/pathway-guides/facilities-management/ [Accessed 4 April 2020].

Sacks, R., Koskela, L., Dave, B.A., Owen, R. (2010a). Interaction of lean and building information modeling in construction. *Journal of Construction Engineering and Management*, 136, 968–980.

Sacks, R., Radosavljevic, M., Barak, R. (2010b). Requirements for building information modeling based lean production management systems for construction. *Automation in Construction*, 19, 641–655.

Sacks, R., Barak, R., Belaciano, B., Gurevich, U., Pikas, E. (2013). KanBIM Workflow Management System: Prototype implementation and field testing. *Lean Construction Journal*, 19–35.

Sacks, R., Eastman, C., Lee, G., Teicholz, P. (2018). *BIM Handbook: A Guide to Building Information Modeling For Owners, Designers, Engineers, Contractors, and Facility Managers*. John Wiley & Sons, Hoboken, NJ.

Schwabe, K., Teizer, J., König, M. (2019). Applying rule-based model-checking to construction site layout planning tasks. *Automation in Construction*, 97, 205–219.

Shou, W., Wang, X., Wang, J., Hou, L., Truijens, M. (2014). Integration of BIM and lean concepts to improve maintenance efficiency: A case study. In *Computing in Civil and Building Engineering*, American Society of Civil Engineers: Reston, VA.

Slack, N., Chambers, S., Johnston, R. (2010). *Operations Management*. Pearson Education, Harlow.

Smith, R.E. (2010). *Prefab Architecture: A Guide to Modular Design and Construction*. John Wiley & Sons, Hoboken, NJ.

Smith, P. (2016). Project cost management with 5D BIM. In *Proceedings of the 29th IPMA World Congress WC2015*. Elsevier Science BV, Westin Playa Bonita.

Sobek II, D.K. and Smalley, A. (2011). *Understanding A3 Thinking: A Critical Component of Toyota's PDCA Management System*. CRC Press, New York.

Sriprasert, E. and Dawood, N. (2003). Multi-constraint information management and visualisation for collaborative planning and control in construction. *Journal of Information Technology in Construction*, 8, 341–366.

Succar, B. (2009). Building information modelling framework: A research and delivery foundation for industry stakeholders. *Automation in Construction*, 18, 357–375.

Succar, B. and Poirier, E. (2020). Lifecycle information transformation and exchange for delivering and managing digital and physical assets. *Automation in Construction*, 112, 103090.

Sydora, C. and Stroulia, E. (2020). Rule-based compliance checking and generative design for building interiors using BIM. *Automation in Construction*, 120, 103368.

Talebi, S. (2014a). Exploring advantages and challenges of adaptation and implementation of BIM in project life cycle. In *2nd BIM International Conference on Challenges to Overcome, BIMForum*. Lisbon.

Talebi, S. (2014b). Rethinking the project development process through use of BIM. In *2nd BIM International Conference on Challenges to Overcome, BIMForum*. Lisbon.

Tauriainen, M., Marttinen, P., Dave, B., Koskela, L. (2016). The effects of BIM and lean construction on design management practices. *Procedia Engineering*, 164, 567–574.

Terreno, S., Asadi, S., Anumba, C. (2019). An exploration of synergies between lean concepts and BIM in FM: A review and directions for future research. *Buildings*, 9, 147.

Tezel, A. and Aziz, Z. (2017). Visual management in highways construction and maintenance in England. *Engineering, Construction and Architectural Management*, 24, 486–513.

Tillmann, P. (2020). Using the Last Planner System to tackle the social aspects of BIM-enabled MEP coordination. *Canadian Journal of Civil Engineering*, 47, 140–152.

Tillmann, P. and Sargent, Z. (2016). Last planner & BIM integration: Lessons from a continuous improvement effort. In *Proceedings of the 24th Annual Conference of the International Group for Lean Construction*, Boston, MA.

Umstot, D. and Fauchier, D. (2017). *Lean Project Delivery: Building Championship Project Teams*. CreateSpace, London.

Vermeulen, D. and Ayoubi, M.E. (2019). Using generative design in construction applications [Online]. Available at: https://www.autodesk.com/autodesk-university/article/Using-Generative-Design-in-Construction-Applications.

Vestermo, A., Murvold, V., Svalestuen, F., Lohne, J., Laedre, O. (2016). BIM-stations: What it is and how it can be used to implement lean principles. In *24th Annual Conference of the International Group for Lean Construction*, Boston, MA.

Volk, R., Stengel, J., Schultmann, F. (2014). Building Information Modeling (BIM) for existing buildings – Literature review and future needs. *Automation in Construction*, 38, 109–127.

Wang, J., Sun, W., Shou, W., Wang, X., Wu, C., Chong, H.-Y., Liu, Y., Sun, C. (2015). Integrating BIM and LiDAR for real-time construction quality control. *Journal of Intelligent & Robotic Systems*, 79, 417–432.

Wang, P., Wu, P., Wang, J., Chi, H.-L., Wang, X. (2018). A critical review of the use of virtual reality in construction engineering education and training. *International Journal of Environmental Research and Public Health*, 15, 1204.

Williams, H. and Duray, R. (2017). *Making IT Lean: Applying Lean Practices to the Work of IT*. CRC Press, Boca Raton, FL.

Womack, J.P., Jones, D.T., Roos, D. (1990). *The Machine that Changed the World*. Rawson Associates, New York.

Wong, J.K.W. and Zhou, J. (2015). Enhancing environmental sustainability over building life cycles through green BIM: A review. *Automation in Construction*, 57, 156–165.

Zaker, R. and Coloma, E. (2018). Virtual reality-integrated workflow in BIM-enabled projects collaboration and design review: A case study. *Visualization in Engineering*, 6(1), 1–15.

Zhang, S., Teizer, J., Lee, J.-K., Eastman, C.M., Venugopal, M. (2013). Building Information Modeling (BIM) and safety: Automatic safety checking of construction models and schedules. *Automation in Construction*, 29, 183–195.

Zhang, S., Sulankivi, K., Kiviniemi, M., Romo, I., Eastman, C.M., Teizer, J. (2015). BIM-based fall hazard identification and prevention in construction safety planning. *Safety Science*, 72, 31–45.

Zhang, J., Long, Y., Lv, S., Xiang, Y. (2016). BIM-enabled modular and industrialized construction in China. *Procedia Engineering*, 145, 1456–1461.

Zou, Y., Kiviniemi, A., Jones, S.W. (2017). A review of risk management through BIM and BIM-related technologies. *Safety Science*, 97, 88–98.

7

Building Information Modeling for Existing Buildings – Deconstruction Planning and Management

Rebekka VOLK
Institute for Industrial Production,
Karlsruhe Institute of Technology, Germany

7.1. Introduction

In recent years, the construction and deconstruction (C&D) sector is digitally developing via so-called building information modeling (BIM). BIM was originally designed to plan and construct new buildings with a shared digital building model. Today, these models not only store building data in a centralized repository and structured way that makes it reusable for different purposes, but also allow for stakeholder cooperation. BIM information can be used to manage, retrofit, renovate or dismantle/ deconstruct existing buildings. Recent trends show the application shift from use in design to application in retrofitting projects (Hübner et al. 2018). But BIM for end-of-life scenarios of buildings is often neglected (Akinade et al. 2017a) despite a “growing awareness of Construction and Demolition (C&D) waste issues” (Charef et al. 2019).

In industrialized countries, there is an enormous stock[1] of existing profane, historic and landmarked buildings (Sanchez and Haas 2018) and old, partly obsolete facilities[2] (e.g. nuclear power plants). However, most of the existing buildings have no BIM (Hossain and Yeoh 2018). In addition, "modifications and deviations of the original building structure, equipment and fittings as well as the deterioration and contamination of buildings are often not well documented or only available in an outdated and unstructured way" (Volk et al. 2018). Thus, the information is not usable for effective deconstruction or disassembly planning, reuse, recycling and waste management purposes. To plan deconstruction measures in existing buildings, "buildings are [currently] audited manually or with stationary laser scans which requires great effort of skilled staff and expensive equipment" (Volk et al. 2018).

The deconstruction and C&D waste management sectors could strongly benefit from chances of digitalization (see Won and Cheng 2017), for example, with respect to documentation of building materials, components, products, layers and coatings, data on deconstructability (joints, bolts), planning of future reuse, recycling or disposal of building components, cost quantification, as well as optimized project and resource management. But BIM "use for assessing the end-of-life impacts is not a common practice" (Akanbi et al. 2019).

This chapter describes the current status of BIM use at the end of a building's lifecycle. Section 7.2 focuses on the state-of-the-art data generation of existing buildings including different data generation schemes, technologies and standards. Section 7.3 reviews the current approaches and developments regarding BIM use in deconstruction. In section 7.4, a summary and an outlook on future R&D directions is given.

1 In EU-28 countries, for example 330.2 million existing residential and non-residential buildings in 2001 cause a high energy demand and GWP [CO_2e] emissions during use and potential secondary raw materials at the end-of-life stage. Calculated by the sum of *Dwellings by type of ownership, type of building and occupancy status* for 2001 in EU-28 countries (except for Germany due to missing values (based on EUROSTAT (2019c), except for Croatia, Malta, Sweden, the United Kingdom, Montenegro, North Macedonia, Serbia, Turkey, Bosnia and Herzegovina). In Germany, ca. 90% of existing buildings were constructed before 2000 (Zensus 2011).

2 Numerous European, North American and Japanese nuclear power plants are at the brink of dismantling (see Volk et al. 2019).

7.2. Data generation for BIM use in existing buildings

7.2.1. *Scan-to-BIM methods*

"Scan-to-BIM is the process of converting 3D reconstructions into building information models (BIM)" (Czerniawskia and Leite 2018). To audit an existing building for deconstruction purposes[3], several methods are possible: They can be classified into non-contact and contact techniques (Volk et al. 2014) and range from manual auditing with checklists to photo or sample taking and from manual to semi-automated and automated methods. Especially in non-contact image-based (photogrammetry, videogrammetry) and range-based (laser scanning, laser measuring) techniques a lot of research has been done recently to support the creation of BIMs for existing buildings (Volk et al. 2014).

Photo- and videogrammetry construct 3D point clouds and 3D models from multiple images. Methods include close-range (<300 m) and aerial (>300 m) photogrammetry. While close-range photogrammetry is more suitable for capturing buildings and infrastructures, aerial methods are applied for survey, mapping and cartography. Both types become indistinct by using drone-mounted cameras, for example, based on DJI drones (DJI 2019), with lower flight altitude. Photogrammetry uses the concept of stereoscopy to detect and extract spatial distances and relations from multiple, partly overlapping images. Factors such as image resolution, lighting conditions and distance from the object influence the 3D model quality. Free and commercial solutions for respective image processing include Autodesk® ReCap™ (Autodesk 2019), DroneDeploy or Pix4D (All3DP 2019).

Laser scanning methods directly capture 3D point clouds and differ in triangulation, time-of-flight and phase-comparison. Furthermore, they can be differentiated into stationary and mobile systems with different technology readiness levels, commercialization statuses and development perspectives. An overview on the type (table, hand scans, mobile version, terrestrial, car-mounted, drone-mounted), the precision (mm) and ranges (m) is given by Boardman and Bryan (2018). While triangulation is suitable for small objects, time-of-flight and phase-comparison are suitable for building data capture. Phase-comparison captures denser point clouds than time-of-flight

3 Hossain and Yeoh (2018) provide an overview on recently applied methods of scan-to-BIM for O&M purposes.

lasers and is thus more adequate for the documentation of landmarks and historic buildings (Boardman and Bryan 2018). Stationary laser scanning systems are commercially available, for example, Leica Panorama Camera (Leica 2019), Z+F IMAGER® 5016 and Z+F IMAGER® 5006i (ZF 2019a, 2019b) or FocusS Laserscanner (Faro 2019a, 2019b) in combination with processing software. Newer developments also provide mobile solutions, for example, ResourceApp with Microsoft Kinect (Volk et al. 2018) in R&D or DotProduct Handscanner (Leica 2019, German Version) as a commercial solution.

After capturing indoor and outdoor laser scanning or photogrammetry point clouds, they are registered (aligned) which is often done via so-called targets or features. Then, the point clouds are cleaned from disturbing elements (e.g. vegetation, furniture, neighboring buildings), segmented and classified so that each point is associated with an object. This is done via criteria such as measured distance, physical location, reflection intensity or color (Volk et al. 2014; Boardman and Bryan 2018) or by annotated images and deep learning approaches (Czerniawski and Leite 2018).

Object detection in the acquired information follows object classification methods autonomously/automatically (Arthur et al. 2018; Irmler et al. 2018), semi-automatically (Lu et al. 2018; Huhnt et al. 2018; Volk et al. 2018) or manually. Furthermore, we can differentiate between object and space classification as well as topology reconstruction, where the latter relates to the derivation of the semantics and relationships between objects (see, e.g., Franz et al. 2017; Czerniawski and Leite 2018; Volk et al. 2018; Huhnt et al. 2018; Lu et al. 2018; Irmler et al. 2018; Tran et al. 2018). Commercial software packages have respective point cloud import, tracing and some automated processing functionality for walls, floors, windows, doors and piping to derive complex 3D models and renderings (Czerniawski and Leite 2018).

Also, these techniques are combined (e.g. Angelini et al. 2017) or complemented by other methods such as infrared thermography, hyperspectral analysis (Amano et al. 2018), ground radar penetration (Hossain and Yeoh 2018), ultrasonics, tags such as RFID (Motamedi et al. 2016), QR or bar codes (Lorenzo et al. 2014) and Internet of Things (IoT) (Teizer et al. 2017; Arthur et al. 2018; Irmler et al. 2018; Shahinmoghadam and Motamedi 2019).

In total, there are many recent approaches to generated as-built BIM from stationary or mobile data capturing methods, also combined with other technologies[4]. However, few studies manage to automatically create IFC objects (Lu et al. 2018) and to verify their results, for example with manual measuring and validation of the building objects' and spaces' proportions, the topological relations between objects and other semantic information (such as texture, surface coatings, material, engineering or quality attributes). Available research and practical approaches seem suitable for simple building structures and larger elements. However, for more complex and irregular shapes like in historical buildings, the conversion is still imprecise or limited to a textured point cloud without semantic information (García-Valldecabres et al. 2016).

7.2.2. *Other methods*

Other methods include automated generation of 3D models based on scanned or digital floor plans/blueprints (e.g. Gimenez et al. 2016) or shape grammar approaches (e.g. Tran et al. 2018). When using 2D plans, the relevant graphical information is extracted and transformed into IFC building elements by automated pattern recognition. The reconstruction is based on the identification of geometry, topology and semantics (Gimenez et al. 2016). A case study shows promising results for walls, openings and spaces in a short processing time (Gimenez et al. 2016).

The shape grammar method uses logical rules to infer, predict and model interior spaces from missing or incomplete data in 3D parametric IFC models (Tran et al. 2018). Furthermore, topological relations (e.g. containment, adjacency and connectivity) are derived. The resulting BIM has high geometric accuracy and rich semantic content.

7.2.3. *Standardized denomination of BIM data elements*

For construction purposes, efforts have been made to standardize BIM structure as well as denomination of building elements/components (entities)

---

4 Best source for information on the topic can be found in the journals *Automation in Construction* and *Advanced Engineering Informatics* and the conferences "International Workshop on Intelligent Computing in Engineering (EG-ICE International Workshop)" and "International Symposium on Automation and Robotics in Construction (ISARC)".

and their properties (attributes[5]) (e.g. see buildingSMART 2019a; BIM Level 2 2019). IFC (EN ISO 16739) is the international standard for this. However, the IFC standard does not address possible values for object types and attributes (Böger et al. 2018) that are required for application of specific BIM functionalities. However, a recent work provides an extension of buildingSMART Data Dictionary that includes appropriate attribute values (Böger et al. 2018).

Regarding interoperatbility of BIM with other software, latest developments focus on BIM-to-FM information interoperability in standardization (e.g. CoBie standard) or research (see e.g. Chen et al. 2018). However, there is still room for improvements for attribute value definition and data mapping engines for BIM (IFC) to connect it more reliably to other (e.g. FM) systems (Böger et al. 2018; Chen et al. 2018).

For deconstruction purposes, no model view definition (MVD) is available to date (buildingSMART 2019b). Instead, for the latest IFC4 version, the Design Transfer View[6] (DTV 1.1) is officially available, while the Quantity Takeoff View[7] (QV 0.1), Product Library View[8] (LV 0.1) and Construction Operations Building Information Exchange[9] (COBie 2.4) are only available as a draft version. However, many studies propose necessary attributes for deconstruction purposes (Akinade et al. 2015; Akinade et al. 2017a; Kühlen 2017; Volk 2017; Hübner 2019). However, standardization processes with respect to BIM and deconstruction functionality are not known to the author.

7.3. BIM use in deconstruction and EOL building stages

7.3.1. *Definitions*

The often synonymously used terms disassembly, decommissioning, reverse engineering and deconstruction[10] have the aim to "eliminate demolition as an end-of-life-option" (Akinade et al. 2017a) through the

5 "Attributes can base on an elementary data type, or they can be relations to other entities". (Böger et al. 2018, p. 308).

6 For advanced geometric and relational representation of spatial and physical components.

7 To estimate and track construction materials and costs.

8 To see manufacturer product information and configurations.

9 To see lifecycle information for maintaining equipment and systems within buildings.

10 In the following, the term "deconstruction" is used.

recovery of reusable materials (Gorgolewski 2006). In the best case, no materials are landfilled – but in reality this is hardly possible (Akinade et al. 2015).

Deconstruction, especially selective deconstruction or dismantling, as a reverse construction process is at least as complex and sophisticated as the construction process, especially because of many undocumented conditions of the building (Hübner et al. 2017; Volk 2017). Similarly, reverse engineering, reverse construction or disassembly is used to describe the removal of components in a reverse construction process, especially for the purpose of reuse (e.g. Sanchez and Haas 2018). The terms decommissioning or dismantling are often used in the context of nuclear dismantling (e.g. Hübner 2019; Volk et al. 2019). It is often defined as the lifecycle stage after the use phase, starting from the shutdown of the building, (infra)structure or facility, until the complete removal to "green field". The term demolition refers to the destruction of a building and infrastructure with the aim to remove it (Volk 2017) and not necessarily with the focus on reuse, separation or recycling of the materials and components.

Further processing steps can be separation, sorting, cleaning, refurbishment or reprocessing, handling and transportation of the building components and products to the new installation, reprocessing or disposal site.

Waste management summarizes the management of unwanted products and residues from (industrial) processes owners want to get rid of. Usually, literature distinguishes between construction and demolition waste. The latter includes products and residues (building components, materials) from disassembly, decommissioning, deconstruction, dismantling and demolition works in buildings and infrastructures. Usually, before deconstruction, a waste management plan has to be developed that documents the whereabouts of all materials, components, products and resulting waste fractions.

End-of-life or "grave" summarizes the last stage in a product's lifecycle. In this lifecycle stage, the function of the product at its location/site is terminated (e.g. a building is removed). If a building component/product or a whole part of a building is reused, then it re-enters the circular economy in a new use phase. The functional reuse of whole building components and products can be performed after separation, decommissioning and further processing steps. It can take place at the same site (e.g. in a replacement building) or at a new site in the installation of a new building or infrastructure. Recycling defines the reuse of building materials, components and products after the destruction of their structure and post-processing to a new component or product made from only or partly secondary raw material (building as a secondary mine or material bank).

7.3.2. *Benefits and impact of BIM deconstruction use case*

Main benefits of BIM for deconstruction planning are basically on an operational level[11], as decisions regarding resource assignment, staff attendance and resource availability information or their schedule and the logistics onsite can be addressed. In deconstruction planning, BIM can be used as a centralized and structured data management. Main benefits lie in the potentials for (1) collaboration and (2) project management functionalities (responsibilities management, planning, performance tracking, monitoring, documentation), (3) the visualization and (4) the automated extraction of the bill-of-quantities/quantity takeoff (list of building materials, components and products) (Won and Cheng 2017) to (5) assign and document possible and suitable reuse, recycling and disposal strategies (waste quantification and management, simulation of end-of-life alternatives, recovery and recycling rates) (Akinade et al. 2017a; Won and Cheng 2017). Furthermore, other operational or management issues, such as (6) cost calculation, (7) handling, transportation and logistics, (8) scaffolding requirements, (9) health and safety considerations (Akinade et al. 2017a) and (10) location-/workspace-based planning (Volk et al. 2017) are mentioned in the literature. In addition, further functionalities could be established such as analysis and optimization functions to (11) maximize reuse or recycling rates (e.g. see Sanchez and Haas 2018), (12) minimize project cost, time, emissions or site disturbance (and associated noise, pollution, vibration) or (13) preserve embodied economic value or embodied energy of materials in a circular economy (e.g. see Akinade et al. 2015; Akinade et al. 2017a; Akanbi et al. 2019).

11 Questions of strategic project planning are rather not addressed by BIM. See Hübner et al. (2017) for a definition of operational and strategic project management and a comprehensive review of current deconstruction project planning approaches.

7.3.3. *Requirements for BIM deconstruction use case*

To plan a deconstruction project with BIM, there are some key requirements that have to be fulfilled beforehand:

1) a pre-existing BIM (see Volk et al. 2014) that has to be created manually, semi-automated or automated in advance (see section 7.2.1);

2) definition of relevant information for deconstruction/disassembly purposes and eventually additional component attributes (MVD) (e.g. restriction matrices, contact matrices for the bolts/connections of elements, locations and workspaces) (see section 7.2.3);

3) standards and unified naming of building elements as it is/was developed for new construction functionalities (see section 7.2.3);

4) interoperability with currently used project management and documentation tools (e.g. MS Excel, MS Project, MS Access, Primavera V6, CORA-CALCOM, ReVK, Siempelkamp[12]) (e.g. see Hübner 2016; Hübner et al. 2018);

5) process maps (e.g. see Won and Cheng 2017), use cases and best practices for application support;

6) training of staff in the C&D sector.

7.3.4. *State-of-the-art deconstruction[13] planning*

7.3.4.1. *Approaches for existing buildings*

In the following, recent approaches are presented and opposed to the possible benefits listed in section 7.3.2.

BIM as a central platform and data repository itself provides basic collaboration and project management functionalities (benefits 1, 2) that can both used for construction and deconstruction purposes.

12 CORACALCOM, ReVK and Siempelkamp Software provide specialized software solution to plan the decommissioning of nuclear facilities, to guarantee the safety and effectiveness of the various processes (Siempelkamp 2019).

13 Synonymously used here for (selective) deconstruction, (selective) disassembly, reverse engineering, reverse construction, decommissioning, dismantling and demolition. See section 7.3.1 for definitions.

Earliest BIM-based works with deconstruction project planning purposes are Cheng and Ma (2013) and Akbarnezhad et al. (2014). They focus on the quantity take-off (benefit 4), the material classification (benefit 5), a coarse cost-based quantity derivation and cost calculation (benefit 6), the required number of hauling trucks as well as their frequency (benefit 7) (Cheng and Ma 2012, 2013) and the accruing material from renovation and deconstruction measures designated for recycling or disposal facilities (Akbarnezhad et al. 2012, 2014).

All approaches mentioned in this section consider the assignment of building components to different possible and suitable reuse, recycling and disposal strategies (benefit 5). Some of the works (Galic et al. 2014; Sanchez and Haas 2018) specifically focus on the relocation and reuse (benefits 11, 13) of building components with BIM. They consider rather coarse structural elements such as steel beams and propose a selective disassembly planning that considers component connections, hosting relations, disassembly levels and other physical constraints such as removal directions to minimize environmental impact and cost.

Newer works focus on the minimization of project time (benefit 12) (Volk 2017; Volk et al. 2018), minimization of total project cost (benefits 6, 12) (Hübner 2019), minimized, near-optimal environmental impact and removal costs (benefit 12) (Sanchez and Haas 2018), minimization of local emissions such as noise, dust and vibrations (benefits 9, 12) (Kühlen 2017) or a multi-objective optimization to optimize cost, delay and recovery rate (benefits 6, 7, 11, 12) (Queheille et al. 2019) based on semantic building models, their bill of quantities and technical constraints to derive deconstruction project planning. Technical constraints consist of, for example, available resource and workspace capacities, suitability of technical equipment and deconstruction ways for building elements and their materials, maximum emission levels and technical/topological precedence relations. For calculation and decision support, they use optimization models such as the (Multi-mode) Resource Project Constraint Scheduling Problem ((M)RCPSP) (Volk 2017; Volk et al. 2018), the Nuclear Dismantling Problem (Hübner 2019), Multi-Objective Optimization Problem (Queheille et al. 2019) and Multi-Attribute Decision Making (Kühlen 2017) to depict multiple independent conflicting economic and environmental objectives.

However, despite very detailed engineering and scheduling approaches, most of these works (Kühlen 2017; Volk 2017; Volk et al. 2018; Hübner 2019; Queheille et al. 2019) use semantic but non-BIM building information. For a comprehensive literature review on previous (selective) deconstruction optimization for semantic, non-BIM models, see Volk (2017) and Hübner et al. (2017)[14].

Different locations or work spaces onsite (benefit 10) are considered by Volk (2017) and Sanchez and Haas (2018) to ensure that reasonable physical spaces are available for the planned deconstruction activities.

Most recent works also include BIM-based visualization (benefit 3) (Marino et al. 2017; Akanbi et al. 2019) and a deconstruction protocol (benefits 4, 5) (Akanbi et al. 2019).

Furthermore, simulation methods in deconstruction planning were considered in Akbarnezhad et al. (2012, 2014), Cheng and Ma (2012, 2013), Volk (2017), Sanchez and Haas (2018) and Hübner (2019) to compare alternative plans or different deconstruction options or the impact of uncertainties. All provide decision support for the "best" option. Only Volk (2017) and Hübner (2019) select a robust schedule for projects under uncertainty, for example, due to incomplete building information. Latest works (Akanbi et al. 2019) provide BIM-based deconstruction simulation functionality. However, model outputs besides the quantified material masses are unclear.

Currently, approaches are missing to cover only scaffolding requirements (benefit 8) and health and safety issues in deconstruction[15] (benefit 9). The other benefits are more or less addressed in research (see Table 7.1).

Currently applied project management software in the deconstruction industry is restricted to general project management software that is mostly not adapted to the specific needs of deconstruction projects (Hübner et al. 2016; Gehring et al. 2021).

14 Hübner et al. 2017 and Hübner 2016 review strategic and operational approaches of building deconstruction or decommissioning project planning and related software.

15 Here, only works for construction projects are available, for example, see Kim et al. (2016).

	Benefits of using BIM for deconstruction purposes	References									
		Cheng and Ma (2012, 2013)	Akbarnezhad et al. (2012, 2014)	Galic et al. (2014)	Sanchez and Haas (2018)	Volk et al. (2018), Volk (2017)	Hübner (2019)	Kühlen (2017)	Queheille et al. (2019)	Akanbi et al. (2019)	Marino et al. (2017)
(1)	Collaboration	(X)	(X)	(X)	(X)	–	–	–	–	(X)	–
(2)	Project management	(X)	(X)	(X)	(X)	X	X	X	X	(X)	–
(3)	Visualization	–	–	–	–	(X)	(X)*	(X)	–	X	X
(4)	Automated extraction of the bill-of-quantities/quantity takeoff	(X)	(X)	(X)	(X)	X	–	X	X	(X)	X
(5)	Documentation of possible/suitable reuse, recycling and disposal strategies	X	X	X	X	X	–	X	X	X	–
(6)	Cost calculation	X	X	–	–	(X)	X	(X)	X	–	–
(7)	Handling, transportation and logistics	X	–	–	–	–	–	–	X	–	–
(8)	Scaffolding requirements	–	–	–	–	–	–	–	–	–	–
(9)	Health and safety	–	–	–	–	(X)	–	X	–	–	–
(10)	Location-/workspace-based planning	–	–	–	X	X	–	–	–	–	–
(11)	Maximize reuse/recycling rates	–	–	X	X	–	–	–	X	X	–
(12)	Minimize project cost, time, emissions or site disturbance	–	–	–	X**	X	X	X	X	–	–
(13)	Preserve embodied economic value or embodied energy	–	–	X	X	–	–	–	–	(X)	–

X: focus of the work; (X): implicitly covered; -: not covered; *: via Gantt charts; **: only near-optimal solution

Table 7.1. *Overview on benefits of using BIM for deconstruction purposes and their covering by the literature*

7.3.4.2. *Approaches for new buildings*

"There has been consensus across literature that design stage is very crucial in construction waste minimization […]. By taking adequate waste minimization strategies during the design stage, about a third of construction waste could be prevented […]" (Ajayi and Oyedele 2017).

DfD strategy	Reference	Used tools and methods	Output data
Durmisevic's knowledge model	Durmisevic (2006)	Fuzzy logic, disassembly sequences	Aggregated score
Building Information Modeling-based Deconstructability Assessment Score (BIM-DAS)	Akinade et al. (2015)	Unweighted checklist	Aggregated score
Adaptive reuse potential (ARP)	Langston et al. (2008)	Multi-criteria sustainability analysis tool	Useful life (in years)
AdaptSTAR	Conejos et al. (2013)	Weighted checklist	Aggregated score
Sequence disassembly planning for buildings (SDPB) method	Sanchez and Haas (2018)	Disassembly sequence structure graph	Optimized disassembly sequence (based on costs and time)

Table 7.2. *Overview on DfD strategies (based on Denis et al. 2018)*

In the literature, several studies focus on design-for-deconstruction or design-for-disassembly (DfD) (Akinade et al. 2015; Akinade et al. 2017a, 2017b; Akinade et al. 2018; Denis et al. 2018; Yeoh et al. 2018; Akanbi et al. 2019; Charef et al. 2019[16]). DfD approaches aim at modifying building design with respect to different criteria that could be helpful with respect to deconstruction, reuse, recycling and waste at the end of a building's lifecycle.

16 See Akanbi et al. (2019) and Table 7.1 for a comprehensive list of previous work.

DfD is seen as a promising design strategy to improve resource efficiency in buildings and facilitate repair, adaptation, and reuse of a building and its elements (Denis et al. 2018). "To facilitate its application in design and construction practice, specific assessment tools are currently being developed" (Denis et al. 2018). As this is not the focus of this article, we refer to Denis et al. (2018) for a review of current DfD approaches: Durmisevic's knowledge model (Durmisevic 2006), Building Information Modeling-based Deconstructability Assessment Score (Akinade et al. 2015), adaptive reuse potential (Langston et al. 2008), AdaptSTAR (Conejos et al. 2013) and the sequence disassembly planning for buildings (SDPB) method (Sanchez and Haas 2018) (see Table 7.2).

7.4. Conclusion

7.4.1. *Summary*

There is a great need and demand for modern and easy-to-use modeling frameworks for building deconstruction in the AEC industry (Marino et al. 2017), to enable circular economy principles and resource efficiency with respect to material, time, cost and energy. Despite many advantages and potential benefits of digitalization of buildings for the purpose of deconstruction, design-for-deconstruction and deconstruction planning at the end of a building's lifecycle is still under development.

Scan-to-BIM or other methods are heavily researched and developed. However, comprehensive research and commercial solutions for the BIM-based deconstruction of existing buildings are not available yet. The most advanced technologies manage to converse point clouds into BIM elements, but the conversion often still is imprecise (García-Valldecabres et al. 2016). Available research and practical approaches seem suitable for simple building structures and larger elements. An intense use of BIM can be seen for complex infrastructures and industrial facilities (e.g. road and rail/metro structures, tunnels, bridges, nuclear power plants) (Ehrbar 2016; Anrijs and Van Steirteghem 2017; Agapaki and Brilakis 2018). However, for more complex and irregular shapes such as in historical buildings, the conversion is imprecise or limited to a textured point cloud without semantic information (García-Valldecabres et al. 2016). For existing buildings' use phase, commercial solutions still focus on heritage and facility management functions (e.g. Faro 2019c).

For technical process and operational models with deconstruction (dismantling, demolition), reuse, recycling and waste management purposes, theoretical project management approaches exist based on semantic, non-BIM building information (e.g. Kühlen 2017; Volk 2017; Volk et al. 2018; Hübner 2019; Queheille et al. 2019). This is because a pre-existing BIM is usually not available for existing buildings that will be deconstructed, and BIM interfaces are complicated or not compatible with other planning software/algorithms due to missing entities and attributes.

BIM-based approaches are still underdeveloped because BIM does not provide comprehensive functionalities and relevant deconstruction attributes on building component and project level. Only the latest works (Akanbi et al. 2019) provide BIM-based deconstruction simulation functionality or simplified (web-based) interfaces (Marino et al. 2017). Recent approaches developed BIM interfaces but neglect the fact of missing BIM models for most of the existing buildings and the related problems. Furthermore, missing standardization, a slow willingness for digitalization and innovation diffusion in the deconstruction sector as well as huge efforts required to digitalize existing buildings for the respective purpose is hampering application.

For new buildings, the main research focuses on "design-for-deconstruction" (DfD) strategies to impact end-of-life stage with BIM from the design stage.

7.4.2. *Outlook*

Besides O&M and facility management purposes, requirements of circular economy, product responsibility, sharing and leasing concepts will change the value of information of existing buildings. Furthermore, improving technologies will facilitate the capturing of existing buildings in a detailed and practical way for the required use cases and functionalities.

However, the current approaches leave room for improvement with respect to comprehensive linkage to BIM, visualization, simulation, optimization and documentation of deconstruction and waste management. For this, IFC objects' standardization and software interfaces, comparative planning with multiple projects or objectives, trade-offs and decision support are required. There is also the need to agree on an appropriate level of detail or granularity for deconstruction functionalities with BIM.

Furthermore, the integration with other digital tools such as GIS, RFID, QR (Lorenzo et al. 2014), mobile augmented reality (García-Valldecabreas et al. 2016; Marino et al. 2017), big data (Bilal et al. 2016) and IoT (Akinade et al. 2017a; Arthur et al. 2018, Dave et al. 2018; Irmler et al. 2018) will be growing. This development might be combined with automated, industrialized disassembly by robots (Marino et al. 2017). In addition, real-time trading in secondary component and raw material markets will be promising to foster a circular economy (also proposed by Akanbi et al. 2019).

7.5. References

Agapaki, E. and Brilakis, I. (2018). State-of-practice on as-is modelling of industrial facilities, in advanced computing strategies for engineering – Part 1. In *25th EG-ICE International Workshop 2018, Lausanne*, Smith, I. and Dommer, B. (eds). Springer.

Ajayi, S.O. and Oyedele, L.O. (2017). Policy imperatives for diverting construction waste from landfill: Experts' recommendations for UK policy expansion. *Journal of Cleaner Production*, 147, 57–65.

Akanbi, L.A., Oyedele, L.O., Motoes, K., Bilal, M., Owolabi, H.A. (2019). Disassembly and Deconstruction Analytics System (D-DAS) for construction in a circular economy. *Journal of Cleaner Production*, 223, 386–396.

Akbarnezhad, A., Ong, K., Chandra, L., Lin, Z. (2012). Economic and environmental assessment of deconstruction strategies using Building Information Modeling. In *Proceedings of Construction Research Congress 2012. Construction Challenges in a Flat World*. West Lafayette, ASCE American Society of Civil Engineers, Reston, VA [Online]. Available at: https://doi.org/10.1061/9780784412329.174.

Akbarnezhad, A., Ong, K.C.G., Chandra, L.R. (2014). Economic and environmental assessment of deconstruction strategies using building information modelling. *Automation in Construction*, 37, 131–144.

Akinade, O.O., Oyedele, L.O., Bilal, M., Ajayi, S.O., Bello, S.A. (2015). Waste minimisation through deconstruction: A BIM based Deconstructability Assessment Score (BIM-DAS). *Resources, Conservation and Recycling*, 105(A), 167–176.

Akinade, O.O., Oyedele, L.O., Omoteso, K., Ajayi, S.O., Looney, J.H. (2017a). BIM-based deconstruction tool: Towards essential functionalities. *International Journal of Sustainable Built Environment*, 6(1), 260–271.

Akinade, O.O., Oyedele, L.O., Ajayi, S.O., Bilal, M., Kabir, O., Kadiri, K.O. (2017b). Design for Deconstruction (DfD): Critical success factors for diverting end-of-life waste from landfills. *Waste Management*, 60, 3–13.

Akinade, O.O., Oyedele, L.O., Ajayi, S.O., Bilal, M., Arawomo, O.O. (2018). Designing out construction waste using BIM technology: Stakeholders' expectations for industry deployment. *Journal of Cleaner Production*, 180, 375–385.

All3DP (2019). 16 best photogrammetry software tools in 2019 (6 are free) [Online]. Available at: https://all3dp.com/1/best-photogrammetry-software/ [Accessed 5 July 2019].

Amano, K., Lou, E.C.W., Edwards, R. (2018). Integration of point cloud data and hyperspectral imaging as a data gathering methodology for refurbishment projects using building information modelling (BIM). *Journal of Facilities Management*, 17(1), 57–75.

Angelini, M.G., Baiocchi, V., Costantino, D., Garzia, F. (2017). Scan to BIM for 3D reconstruction of the papal basilica of Saint Francis in Assisi in Italy. ISPRS – International Archives of the Photogrammetry, Remote Sensing and Spatial Information Sciences, XLII-5/W1, 47–54.

Anrijs, J. and Van Steirteghem, J. (2017). Integrated BIM for failure-free retrofitting of a motorway tunnel. In *IABSE Conference: Creativity and Collaboration*. Bath [Online]. Available at: https://structurae.net/en/literature/conference-paper/integrated-bim-for-failure-free-retrofitting-of-a-motorway-tunnel. DOI:10.2749/222137817821233134.

Arthur, S., Li, H., Lark, R. (2018). The emulation and simulation of Internet of Things devices for building information modelling (BIM), in advanced computing strategies for engineering – Part 2. In *25th EG-ICE international Workshop 2018*, Smith, I., Domer, B. (eds). Springer, Lausanne [Online]. Available at: https://structurae.net/en/literature/conference-paper/integrated-bim-for-failure-free-retrofitting-of-a-motorway-tunnel DOI:10.2749/222137817821233134.

Autodesk (2019). Photogrammetry software [Online] Available at: www.autodesk.com/solutions/photogrammetry-software [Accessed 5 July 2019].

Bilal, M., Oyedele, L.O., Qadir, J., Munir, K., Ajayi, S.O., Akinade, O.O., Owolabi, H.A., Alaka, H.A., Pasha, M. (2016). Big data in the construction industry: A review of present status, opportunities, and future trends. *Adv. Eng. Inform.*, 30, 500–521.

BIM Level 2 (2019). Available at: http://bim-level2.org/en/ [Accessed 5 July 2019].

Boardman, C. and Bryan, P. (2018). *3D Laser Scanning for Heritage – Advice and Guidance on the Use of Laser Scanning in Archaeology and Architecture.* Historic England, Swindon.

Böger, L., Huhnt, W., Wernik, S. (2018). Standardized names of object types and attributes as basis for cooperation between planning and building. In *Advanced Computing Strategies for Engineering – Part 2, 25th EG-ICE International Workshop 2018*, Smith, I., Domer, B. (eds). Springer, Lausanne.

buildingSMART (2019a). buildingSMART International [Online]. Available at: https://www.buildingSMART.org [Accessed 5 July 2019].

buildingSMART (2019b). MVD Database [Online]. Avalaible at: https://technical.buildingSMART.org/standards/mvd/mvd-database/ [Accessed 9 July 2019].

Charef, R., Alaka, H., Ganjian, E. (2019). A BIM-based theoretical framework for the integration of the asset end-of-life phase. *IOP Conference Series: Earth and Environmental Science*, 225(1), article no. 012067 [Online]. Available at: https://pureportal.coventry.ac.uk/en/publications/a-bim-based-theoretical-framework-for-the-integration-of-the-asse.

Chen, W., Chen, K., Cheng, J.C.P. (2018). Towards an ontology-based approach for information interoperability between BIM and facility management. In *Advanced Computing Strategies for Engineering – Part 2, 25th EG-ICE International Workshop 2018*, Smith, I., Domer, B. (eds). Springer, Lausanne.

Cheng, J.C.P. and Ma, L.Y.H. (2012). A BIM-based system for demolition and renovation waste quantification and planning. In *Proceedings of the 14th International Conference on Computing in Civil and Building Engineering (ICCCBE 2012)*, Moscow.

Cheng, J.C.P. and Ma, L.Y.H. (2013). A BIM-based system for demolition and renovation waste estimation and planning. *Waste Management*, 33(6), 1539–1551.

Conejos, S., Langston, C., Smith, J. (2013). AdaptSTAR model: A climate-friendly strategy to promote built environment sustainability. *Habitat Int.*, 37, 95–103.

Czerniawski, T. and Leite, F. (2018). 3DFacilities: Annotated 3D reconstructions of building facilities. In *Advanced Computing Strategies for Engineering – Part 1., 25th EG-ICE International Workshop 2018*, Smith, I., Domer, B. (eds). Springer, Lausanne.

Dave, B., Buda, A., Nurminen, A., Främling, K. (2018). A framework for integrating BIM and IoT through open standards. *Automation in Construction*, 95, 35–45.

Denis, F., Vandervaeren, C., Temmerman, N.D. (2018). Using network analysis and BIM to quantify the impact of Design for Disassembly. *Buildings*, 8(8), 113.

DJI (2019). The world leader in camera drones/quadcopters [Online]. Available at: https://www.dji.com [Accessed 5 July 2019].

Durmisevic, E. (2006). Transformable building structures. Design for disassembly as a way to introduce sustainable engineering to building design & construction. Thesis, Technische Universiteit Delft, Delft.

Ehrbar, H. (2016). Building Information Modelling – A new tool for the successful implementation of major projects of German railways/Building Information Modelling – Ein neues Werkzeug zur erfolgreichen Realisierung von Großprojekt en der Deutschen Bahn. *Geomechanics and Tunnelling*, 9(6), 659–673.

EUROSTAT (2019). Building permits – Annual data [Online]. Available at: https://ec.europa.eu/eurostat/web/products-datasets/-/sts_cobp_a [Accessed 26 April 2019].

Faro (2019a). FARO Focus: Laserscanner-Serie – Die perfekten Werkzeuge für 3D-Dokumentation in den Bereichen Architektur/Baugewerbe/Konstruktion, Öffentliche Sicherheit/Forensik und Produktdesign [Online]. Available at: https://www.faro.com/de-de/produkte/bausektor-bim-cim/faro-focus/ [Accessed 22 May 2019].

Faro (2019b). Solutions for facility management – Scanning for the facility lifecycle [Online]. Available at: https://constructionbim.faro.com/us/facility-management/ [Accessed 22 May 2019].

Faro (2019c). Construction BIM – Informed lifecycle [Online]. Available at: https://constructionbim.faro.com/us/ [Accessed 22 May 2019].

Franz, S., Irmler, R., Rüppel, U. (2017). Collaborative BIM reconstruction on mobile consumer devices. In *Digital Proceedings of the 24th International Workshop of the European Group for Intelligent Computing in Engineering*, Koch, C. (ed.). European Group for Intelligent Computing in Engineering (EG-ICE). Nottingham, 10–12 July 2017 [Online]. Available at: https://tubiblio.ulb.tu-darmstadt.de/88504/.

Galic, M., Dolacek-Alduk, Z., Cerovecki, A., Glick, D., Abramovic, M. (2014). BIM in planning deconstruction projects, eWork and eBusiness in architecture, engineering and construction. In *Proceedings of the 10th European Conference on Product and Process Modelling, ECPPM 2014*, DOI: 10.1201/b17396-17, Taylor & Francis, London [Online]. Available at: https://www.taylorfrancis.com/chapters/mono/10.1201/b17396-17/human-requirements-factors-ardeshir-mahdavi-bob-martens-raimar-scherer.

García-Valldecabres, J., Pellicer, E., Jordan-Palomar, I. (2016). BIM scientific literature review for existing buildings and a theoretical method: Proposal for heritage data management using HBIM. In *Construction Research Congress 2016: Old and New Construction Technologies Converge in Historic San Juan*, Perdomo-Rivera, J.L., Gonzáles-Quevedo, A., López del Puerto, C., Maldonado-Fortunet, F., Molina-Bas, O.I. (eds). Curran Associates Inc., New York, and ASCE American Society of Civil Engineers, Reston, VA [Online]. Available at: https://ascelibrary.org/doi/10.1061/9780784479827.222.

Gehring, M., Winkler, F., Volk, R., Schultmann, F. (2021). Projektmanagementsoftware und Scheduling: Aktuelle Bestandsaufnahme von Funktionalitäten und Identifikation von Potenzialen. Karlsruher Institut für Technologie (KIT). doi:10.5445/IR/1000138068.

Gimenez, L., Robert, S., Suard, F., Zreik, K. (2016). Automatic reconstruction of 3D building models from scanned 2D floor plans. *Automation in Construction*, 63, 48–56.

Gorgolewski, M. (2006). The implications of reuse and recycling for the design of steel buildings, *Can. J. Civ. Eng.*, 33(4), 489–496.

Hossain, M.A. and Yeoh, J.K.W. (2018). BIM for existing buildings: Potential opportunities and barriers. In *3rd International Conference on Building Materials and Construction, ICBMC 2018*, Nha Trang.

Hübner, F. (2016). Software zur Unterstützung von Abbrucharbeiten – Welche Programme werden tatsächlich verwendet? Logiciels pour les projets de déconstruction – Quels programmes sont réellement utilisés ? *Abbruch aktuell*, 3, 14–15.

Hübner, F. (2019). Planung und Modellierung des Rückbaus kerntechnischer Anlagen unter der Berücksichtigung von Unsicherheiten – Ein Beispiel zur Planung von Großprojekten. Dissertation. KIT Scientific Publishing. doi:10.5445/KSP/1000091848.

Hübner, F., Volk, R., Kühlen, A., Schultmann, F. (2017). Review of project planning methods for deconstruction projects of buildings. *Built Environment Project and Asset Management*, 7(2), 212–226.

Hübner, F., Volk, R., Secer, O., Kühn, D., Sahre, P., Knappik, R., Schultmann, F., Gentes, S., von Both, P. (2018). Modellentwicklung eines ganzheitlichen Projektmanagementsystems für kerntechnische Rückbauprojekte (MogaMaR): Schlussbericht des Forschungsvorhabens/Model development of an integrated project management system for nuclear decommissioning projects (MogaMaR). Final report, KIT Scientific Publishing, Karlsruhe.

Hübner, F., Gerhards, P., Stürck, C., Volk, R. (2021). Solving the nuclear dismantling project scheduling problem by combining mixed-integer and constraint programming techniques and metaheuristics. *Journal of Scheduling*, 24(3), 269–290.

Huhnt, W., Hartmann, T., Suter, G. (2018). Space classification from point clouds of indoor environments based on reconstructed topology. In *Advanced Computing Strategies for Engineering – Part 1, 25th EG-ICE International Workshop 2018*, Smith, I., Domer, B. (eds). Springer, Lausanne.

Irmler, R., Franz, S., Rüppel, U. (2018). IoT – An opportunity for autonomous BIM reconstruction. In *Advanced Computing Strategies for Engineering – Part 1, 25th EG-ICE International Workshop 2018*, Smith, I., Domer, B. (eds). Springer, Lausanne.

Kim, K., Cho, Y., Zhang, S. (2016). Integrating work sequences and temporary structures into safety planning: Automated scaffolding-related safety hazard identification and prevention in BIM. *Automation in Construction*, 70, 128–142.

Kühlen, A. (2017). Techno-economic and environmentally conscious deconstruction project planning and decision support (TEE-D-Plan). Thesis, Institut de technologie de Karlsruhe (KIT) [Online]. Available at: https://doi.org/10.5445/IR/1000069327.

Langston, C., Wong, F.K.W., Hui, E.C.M., Shen, L.-Y. (2008). Strategic assessment of building adaptive reuse opportunities in Hong Kong. *Build. Environ.*, 43, 1709–1718.

Leica (2019). Leica Cyclone 3D Point Cloud Processing Software – A total point cloud delivery solution, Leica Geosystems [Online] Available at: https://leica-geosystems.com/en-in/products/laser-scanners/software/leica-cyclone [Accessed 22 May 2019].

Lorenzo, T.M., Benedetta, B., Manuele, C., Davide, T. (2014). BIM and QR-code: A synergic application in construction site management. *Procedia Engineering*, 85, 520–528.

Lu, Q., Lee, S., Chen, L. (2018). Image-driven fuzzy-based system to construct as-is IFC BIM objects. *Automation in Construction*, 92, 68–87.

Marino, E., Spini, F., Paoluzzi, A., Salvati, D., Christian Vadalà, C., Bottaro, A., Vicentino, M. (2017). Modeling semantics for building deconstruction. In *Proceedings of the 12th International Joint Conference on Computer Vision, Imaging and Computer Graphics Theory and Applications, GRAPP*, Porto.

Motamedi, A., Soltani, M.M., Setayeshgar, S., Hammad, A. (2016). Extending IFC to incorporate information of RFID tags attached to building elements. *Advanced Engineering Informatics*, 30(1), 39–53.

Queheille, E., Taillandier, F., Saiyouri, N. (2019). Optimization of strategy planning for building deconstruction. *Automation in Construction*, 98, 236–247.

Sanchez, B. and Haas, C. (2018). A novel selective disassembly sequence planning method for adaptive reuse of buildings. *Journal of Cleaner Production*, 183, 998–1010.

Shahinmoghadam, M. and Motamedi, A. (2019). Review of BIM-centred IoT deployment: State of the art, opportunities, and challenges. In *36th International Symposium on Automation and Robotics in Construction (ISARC2019)*, Banff.

Siempelkamp (2019). Software solutions – Modern solutions for complex technical challenges, Siempelkamp Ingeneieursgesellschaft mbH [Online]. Available at: http://www.siempelkamp-nis.com/index.php?id=608&L=1 [Accessed 22 May 2019].

Teizer, J., Wolf, M., Golovina, O., Perschewski, M., Propach, M., Neges, M., König, M. (2017). Internet of Things (IoT) for integrating environmental and localization data in Building Information Modeling (BIM). In *ISARC: Proceedings of the International Symposium on Automation and Robotics in Construction*, Taipei. DOI: 10.22260/ISARC2017/0084.

Tran, H., Khoshelham, K., Kealy, A., Díaz Vilarino, L. (2018). Shape grammar approach to 3d modeling of indoor environments using point clouds. *Journal of Computing in Civil Engineering*, 33(1).

Volk, R. (2017). Proactive-reactive, robust scheduling and capacity planning of deconstruction projects under uncertainty. *Production and Energy Series 20*, Institut de technologie de Karlsruhe, Institut de la production industrielle; KIT Scientific Publishing, Karlsruhe.

Volk, R., Stengel, J., Schultmann, F. (2014). Building Information Modeling (BIM) for existing buildings – Literature review and future needs. *Automation in Construction*, 38, 109–127.

Volk, R., Luu, T.H., Mueller-Roemer, J.S., Sevilmis, N., Schultmann, F. (2018). Deconstruction project planning of existing buildings based on automated acquisition and reconstruction of building information. *Automation in Construction*, 91, 226–245.

Volk, R., Hübner, F., Hünlich, T., Schultmann, F. (2019). The future of nuclear decommissioning – A worldwide market potential study. *Energy Policy*, 124, 226–261.

Won, J. and Cheng J.C.P. (2017). Identifying potential opportunities of building information modeling for construction and demolition waste management and minimization. *Automation in Construction*, 79, 3–18.

Yeoh, J.K.W., Lee, S.L., Ong, K.C.G. (2018). Development of a BIM-based framework to support the analysis of "design-for-disassembly" strategies. In *Construction Research Congress 2018: Sustainable Design and Construction and Education – Selected Papers*, Orleans.

Zensus (2011). Endgültige Ergebnisse: Gebäude- und Wohnungsbestand in Deutschland/Résultats finaux : bâtiments et unités résidentiels en Allemagne, Statistische Ämter des Bundes und der Länder.

ZF (2019a). Z+F IMAGER® 5006i: Die neue Art zu scannen – Hochpräzise, schnell, zuverlässig und flexibel, Zoller+Fröhlich [Online]. Available at: https://www.zf-usa.com/zf-3d-laser-scanners-zf-laser-scanners/ [Accessed 22 May 2019].

ZF (2019b). Z+F 3D Laser Scanners, Zoller+Fröhlich [Online]. Available at: https://www.zf-usa.com/zf-3d-laser-scanners-zf-laser-scanners/ [Accessed 22 May 2019].

8

BIM, GIS: Complementarity and Convergence

Hervé HALBOUT[1,2], François ROBIDA[3,2] and Mojgan A. JADIDI[4]
[1]*HALBOUT Consultants, Caen, France*
[2]*Projet MINnD, Paris, France*
[3]*BRGM : Service géologique national français, Orléans, France*
[4]*Lassonde School of Engineering, York University, Toronto, Canada*

8.1. BIM and GIS

8.1.1. *Definitions*

An information system is used to store, organize and structure data in digital database management systems (DBMS), to process and analyze data for the sharing and distribution of information and knowledge. Generally, data substantiate information. Geospatial data are spatial data that are geolocated relative to Earth (with geographical coordinates), making it possible to produce geographical knowledge useful for planning and analysis of our environment, infrastructures and cities. Geographic information systems (GIS) are therefore information systems (IS), to which a geospatial component (G) is added. It includes databases (with georeferenced data or location information), hardware infrastructures (servers, computers, operating system, software and business applications, etc.), a structural organization (management, service, mission, etc.), human organization (administrator, data scientist, engineers, cartographer, etc.) and overall

Building Information Modeling,

governance (data management) (Goodchild et al. 2007). This requires implementation of specific expertise for production, management, usage, amendment, analysis and mapping of data, as well as making the data available to various users through web portals, clouds, PDF, etc.

Building information modeling (BIM) is a collaborative process between different professions linked to the lifecycle of a building or linear infrastructure (road, rail, bridge, underground networks, etc.). It is based on the implementation and utilization of digital models, a virtual 3D representation of as-built structures – what is going to be built and what is built. A digital model produced as part of a BIM process allows users to visualize, simulate, evolve, operate and maintain buildings and infrastructures from the design and then construction phases to their maintenance and operation. The main advantage of such digital models is their link with tabular databases, allowing the association of description for each component in the digital models.

Indeed, there are significant synergies between BIM and GIS to integrate and work together in many urban infrastructure and building applications. However, interoperability of these two systems is a main challenge due to inconsistent data formats, semantics definitions, and dissimilar data models. Consider that interoperability is the ability of proprietary or open-source business software/applications (developed by digital services companies) to communicate with each other through their specific proprietary format. There are therefore different levels of interoperability of BIM and GIS which will be described in section 8.4.

8.1.2. *GIS, as a technical and organizational tool*

For GIS users, there are different levels between data production and its final use, where each stage is dedicated to a particular initial usage. Data producers create data for their immediate business need. They are responsible for their production, updating and dissemination of data. These datasets can be used by other businesses or organizations (based on the principle of sharing and disseminating), which potentially overlay data with their own business data for analysis purposes. Therefore, this supply chain of data production, sharing and disseminating is an essential element of GIS. It can be extended to an end user (public) who will, for example, simply visualize a result in cartographic form using a web mapping application.

8.1.3. *GIS, a powerful land information management tool*

GIS allows extensive and interactive visualizations by overlaying multiple layers of information (vector and raster data). This includes a representation of real-world phenomena at a given period through a cartographic interpretation of structuring elements, such as roads, rails, bridges, buildings. Except for raster data (orthophotographs, satellite images), vector data are a cartographic representation of an interpreted or simplified version of reality through a graphic symbology (a highway is represented by a thick red line, for example). However, the real strengths of GIS are based on the semantic information associated with these representations, primarily the descriptive attribute data associated with each element, to be characterized and managed in DBMS. These databases will allow the analysis of the real world and the decision support for its development. In addition, each map layer also includes metadata, which is important information in terms of data traceability and lineage. GIS have evolved significantly over time, moving from the field of academic research to land information management system in public and private sectors as presented in Figure 8.1.

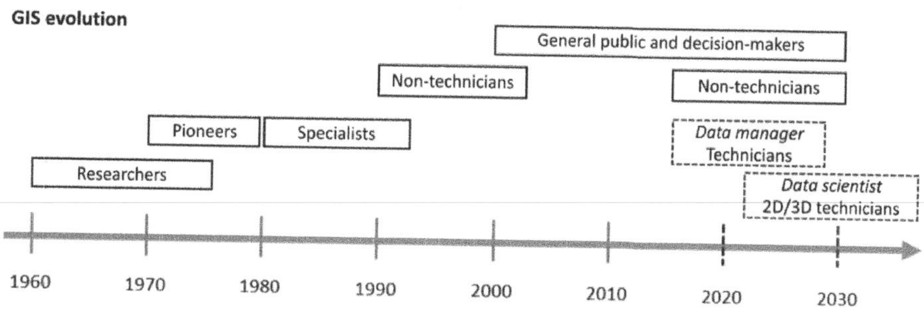

Figure 8.1. *GIS evolution since 1960 (adapted from Bordin and Gimenez 2017)*

Although GIS should not be presented as conceptual components here, there is one area that should be revisited regularly: the data (geospatial or otherwise) that are the main benefactor for any GIS system. After an initial period (1980-1990s), which focused on software tools by publishers and their cartographic representation functionalities, it is the essential component of data that has finally taken the lead in geospatial communities. GIS administrators and geomatics engineers/scientists have understood the necessity of working on this matter, over time, as the main element of their professional activities are focused on data acquisition, production, management and sharing. The structuring of information has become the key word in both relational and non-relational DBMS, and this is what allows easy exchanges between competitive software. The world of GIS has played a very important role in this subject and is largely gaining the benefits. This is one of their strengths that data have become today, in the world of GIS, a real "common good".

Figure 8.2 represents some of the business applications and processes that are concerned with and impacted by the use of a GIS. This diagram does not consider certain business applications, which are not directly producers of geospatial data, such as finance, human resources and social media. However, their production of business data may be cross-referenced with geospatial data.

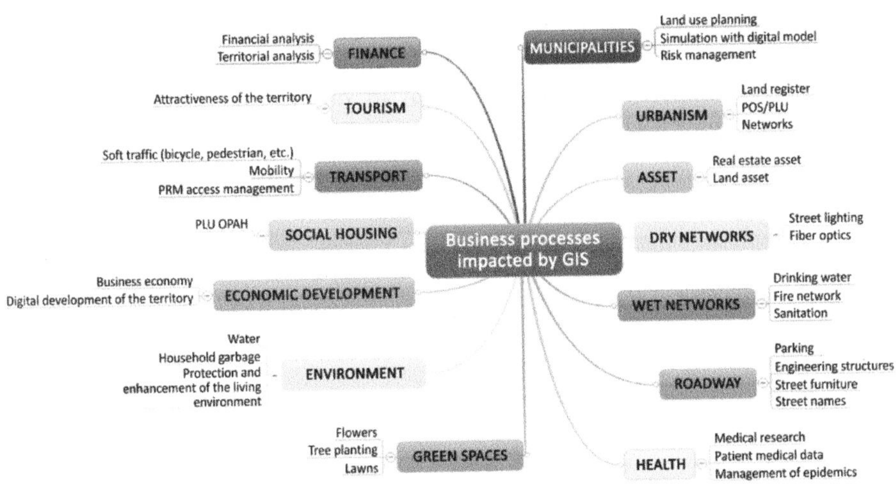

Figure 8.2. *The role of GIS in business intelligence and applications*

In order to produce a cartographic representation (a map) of a territory, it is essential to have geographic data. Software tools are still necessary to help in creating, visualizing, analyzing and sharing geospatial data. GIS software and applications have become mature, meeting the needs expressed by professionals. The use of extract, transfer, and load (ETL) software has also made it possible to a certain degree of interoperability for multiple data sources to better connect in the numerous exchange formats and to automate repetitive tasks related to data management (extraction, topological controls, format transformations, etc.). The advent of spatial servers has made it possible to achieve a form of interoperability in the use of data, by freeing oneself from proprietary software formats. Geospatial information usually represented in a proprietary vector format (editor) is now stored directly in a spatial DBMS (SDBMS) or object-oriented databases. In addition, the development of web technologies and their standardization by the Open Geospatial Consortium (OGC) has led to a more flexible, easier access to the mapping and GIS.

Moreover, the consideration of the vertical dimension of the different components of the real world (subsoil, ground, above ground) leads today's GIS to evolve toward a 3D modeling of geospatial data. The 3D visualization thus fundamentally changes the perception that we can have of the real world, while the geospatial data that represent it in 2D or 3D remains identical in a virtual world. A third coordinate is simply added by a height/altitude Z component. However, not all mapping software is able to manage 3D data and modeling. Specific formats have appeared, proposed both by data providers (proprietary formats) and OGC (City Geography Markup Language [CityGML] format, for example). This 3D representation of the real world is different from a 3D digital model of a building, as it is proposed today in the BIM approach.

8.1.4. *BIM, a powerful asset management tool*

BIM has also undergone its own evolution, different from that of GIS. There are two different communities and two different professions, evolved in their own way, and both are involved in land use planning and are urban data providers. While GIS are tools for managing and analyzing the real world, adapted to scales ranging from a city district to a region, a state or even a global representation, BIM models are more specifically the digital models of built infrastructure (buildings, roads, networks, etc.). BIM allows more targeted and precise management of these infrastructures at a structural element scale, the overall scale being much more limited but focused. Its main focus on buildings makes it a very useful management tool today for the built heritage, facility management and maintenance, connected and smart buildings in conjunction with dedicated business applications. Figure 8.3 represents the current challenges of BIM and digital models, an inseparable representation of the collaborative process.

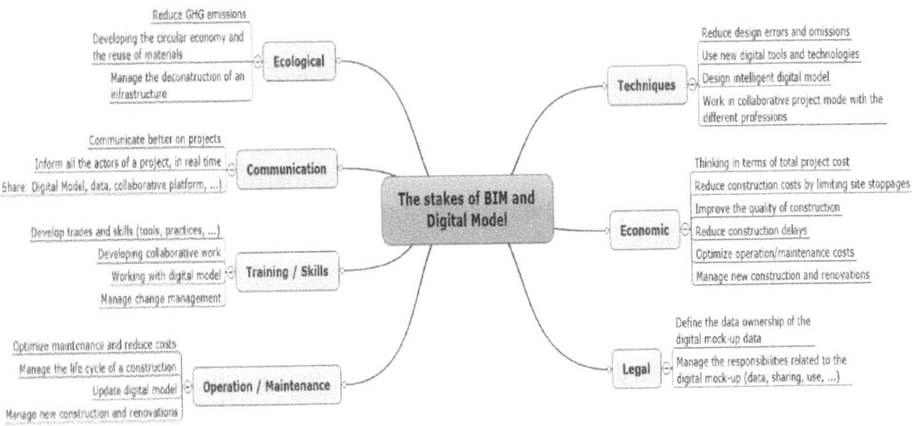

Figure 8.3. *The main challenges of BIM and digital models*

8.2. BIM and GIS: Complementarity/convergence/digital continuity

8.2.1. *Analogies between GIS and BIM*

The development of GIS and BIM have been carried out separately and without any consultation among different cross-domain professions, which could allow an evaluation of potential complementarities – both from a methodological and tools development/usage point of view. This can be explained by different business cultures. GIS was originally developed to "map" existing features of the environment (natural or built environment), where BIM is based on the description of the processes starting from the design and construction of man-made objects (e.g. computer-aided design [CAD] models). The initial users were geographers and natural scientists for GIS, while architects and designers for BIM have different cultures and scientific tools. This explains that the two disciplines were originally developed separately.

Table 8.1 summarizes the most common analogies working in these two fields. We have deliberately separated the 2D GIS from the 3D GIS because if the former is quite mature in the geospatial profession, the latter still struggles to be used widely by professionals. However, we did not try to compare CAD and BIM, since we represent an analogy rather than a difference.

	2D GIS	3D GIS	BIM
Visualization (web and viewer)	Yes (web)	Yes (web)	Yes (viewer/server)
Federative approach (transversal)	Yes	Yes	Yes
Evolving business skills	Yes	Yes	Yes
Exchange formats	Yes (SHP/GeoJSon/XML)	Yes (GML)	Yes (IFC)
Interoperability	Yes (100%)	Partial (50%)	Partial (50%)
Open-source tools (cartography, DBMS, etc.)	Yes (PostGre/PostGis)	Yes (PostGreSQL)	In progress
Data, as a major issue	Yes	Yes	Yes
Different scales of work	Yes (territory)	Yes (territory/district)	Yes (building)
Level of detail or development (LoD)	No	Yes (LoD)	Yes (LoD)
Cataloguing data + metadata (European directive INSPIRE)	Yes	Yes	No
Exchange platforms (sharing/mutualization)	Yes	No	Yes
Exchange platforms (collaboration)	No	No	Yes
Allows for analysis and decision support	Yes	Yes	Yes
Allows operation and maintenance	No	No	Yes

Table 8.1. *Summary of analogies between 2D GIS, 3D GIS and BIM*

8.2.2. *Scale complementarity of GIS and BIM*

There is both a complementarity of scale and a territorial continuity between BIM and GIS. This complementarity of scale is shown in Figure 8.4. Whether it is a building or a linear infrastructure (civil engineering: road, rail, bridge, networks, etc.) treated in the BIM process and the 3D digital model, they are integrated into a larger territory, managed by the 2D and 3D GIS. Whether with BIM or GIS, there are associated descriptive data, and the difference is rather in the field precision of the data.

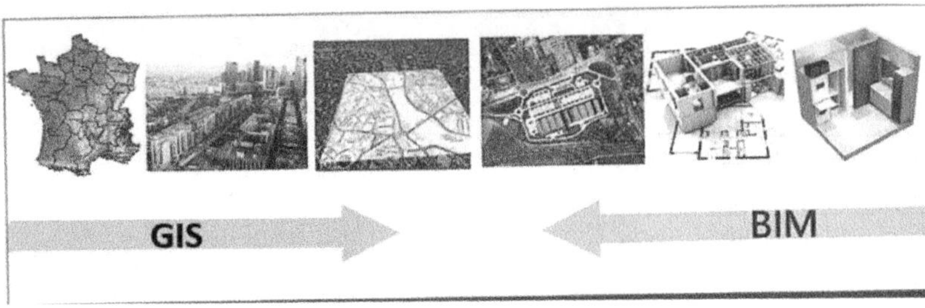

Figure 8.4. *Scale complementarity of GIS and BIM (adapted from Bordin and Gimenez 2017)*

8.2.3. *Complementarity of (geo)localization*

The second complementarity concerns geolocation. A building is not a simple architectural construction which can be located anywhere on the globe; what counts is the construction itself, its geolocation and its content. It is naturally georeferenced in the space, such as the cartographic representation of GIS. Moreover, in GIS, the 2D representation of a building in the form of a polygon or 3D representation in the form of a cube is usefully complemented by the constituent elements of the construction in BIM data.

8.2.4. *Data complementarity*

Beyond the cartographic/architectural representation, BIM and GIS data have an essential commonality and are complementary together. The main evolutions of the software tools for digital models in BIM processes have focused on 3D modeling in object-oriented mode and the association to each object of specific attribute information managed in a database. We are no longer in a simple visual representation as 3D, but as with GIS, rather in the position to query, analyze, interpret and make decisions. Table 8.1 represents a synthesis of the commonality and complementariness between BIM and 3D GIS coupled with the interactive and immersive fields of augmented reality (AR) and virtual reality (VR), which are increasingly using digital models of physical world. This creates a new era of usage of BIM and 3D GIS technologies for simplifying field work, detecting clashes, increasing safety, gaining enhanced and data driven insights as well as reducing errors during and after construction. This also became a new pool of research and development for many startup companies joining the world-known software suites such as Autodesk, ESRI and Bentley for AR and VR (XR) app development for construction automation and facility managements.

8.3. Convergence of formats

8.3.1. *The emergence of GIS standards and the role of OGC*

The development of GIS tools started in the late 1960s by Canadian geographer Dr. Roger Tomlinson while working for the Canadian government – a geographic database still being used today by municipalities across Canada for land planning and integrating multiple sources of geospatial data (Goodchild 2010). Indeed, the management of the link between the geometries, essentially 2D, and the attributes ("semantics") of the objects has constituted a basic function of GIS. In the 1980s, GIS software was developed both by software vendors and large public agencies. However, the integration of multiple sources of data with different natures, types, and format made it difficult and costly for users to transfer data between systems. The need to ensure interoperability between data and tools from different public and private sectors quickly emerged.

Consequently, the Open GIS Consortium, Inc. (OGC) (now called the Open Geospatial Consortium) was launched in 1994. As an international non-profit organization, the consortium brings together the players involved in the development and promotion of open standards guaranteeing interoperability in the field of geomatics and geospatial information through cooperation between developers, suppliers and users. The OGC has thus grown from 20 members in 1994 to 520 in 2020, bringing together government, academic and private sectors and organizations. Although founded in the United States, the OGC now has more members across Europe than elsewhere (Reichardt and Robida 2019).

The OGC established a consensus process among its members to develop specifications to produce all relevant open standards. This process is addressed by taking into consideration the interoperability needs expressed by different communities (meteorology, aviation, geosciences, etc.) through technical domain working groups (Domain Working Group [DWG]). It is based on experiments (testbeds, pilots, Interoperability Experiments) to validate existing solutions or identify challenges. The interested experts collaboratively write the specifications of the standards through working groups (Standards Working Group [SWGs]) according to established best practices. The adoption of the standards is then recommended after testing in open and commercial software. Once the standards have been adopted, a certification process allows developers to have their implementation validated in their software product. Today, OGC standards are integrated into hundreds of open and commercial software products.

The OGC has a very close relationship with ISO/TC211[1] (Geographic Information/Geomatics), which has "transformed" the main OGC standards into ISO standards.

To ensure the consistency of the Internet and web ecosystems, OGC has partnered with many other standards development organizations and industry associations[2]. They work closely with OGC on a wide range of topics such as integration of indoor/outdoor locations, sensor fusion, urban modeling, location-based marketing, aviation, meteorology, internet of things, points of interest and the semantic web. Industrial or technical organizations thus seek to guarantee that their data will be accessible to other communities through OGC standards, and that they can, in return, benefit from the data of other communities. Today, OGC standards are key elements of geospatial communication interfaces, encodings and best practices for sensor networks, location services, terrestrial imagery networks, climate models, climate management programs, disasters and national spatial data infrastructures around the world. Since the beginning of the 2000s, the emergence of spatial data infrastructures (SDIs) deployed on a regional, national, continental or international scale has also benefited from OGC standards and encouraged their deployment (Nebert 2001). In Europe, the INSPIRE law[3], which defines the rules for building a European spatial data infrastructure, is mainly based on OGC standards.

8.3.2. *OGC standards*

The interoperability made possible by OGC is mainly based on a web architecture through the implementation of standard "web services". The best known of these, the web map service (WMS) is a standard communication protocol that allows georeferenced data maps to be obtained

1 Available at: https://committee.iso.org/home/tc211.

2 W3C, OASIS, IETF, bSI (buildingSMART International), WMO (World Meteorological Organisation), etc.

3 Available at: https://inspire.ec.europa.eu/inspire-directive/2.

from different data servers that are distributed over the web (De la Beaujardiere 2002). This makes it possible to set up a network of map servers from which clients can build interactive maps offering the superposition of data of different origins and natures. Changes in projection systems are taken into account by the service.

Other services allow access not to a cartographic representation but to the data itself, such as the web feature service (WFS) or the web coverage service (WCS).

The catalog service for the web (CSW) cataloging protocol makes it possible to expose and search the various services for accessing the data or their representations made available by geospatial data distributors.

In addition to these web service specifications, OGC also offers geography markup language (GML)-based data exchange formats. GML is a language derived from XML for encoding, manipulating, and exchanging geographic data. The GML consists of a set of XML schemas that define an open format for the exchange of geographic data. The GML language is used to describe geographic objects, projection systems, geometry (1D, 2D or 3D, vector or raster), topology, time, units of measurement and attributes of geographic objects.

Based on GML, it is possible to build specific data models for specialized areas ("community standards"), such as CityGML for the city, WaterML for hydrology or GeoSciML for geology.

Finally, the OGC proposes a very generic model to describe observations and measurements (O&M for Observations and Measurements), which are based on a series of specifications to standardize the access of sensor data, whether it is a thermometer, a probe measuring the depth of water in a borehole or satellite observation. This model is also applicable to observations made by a "human" sensor (biodiversity, geology, etc.). O&M is adopted in many environmental disciplines.

The standards for accessing sensor data are grouped together in the sensor web enablement. They make it possible to build monitoring and alert systems based on heterogeneous sensors distributed and managed by different actors, accessible through the web.

Although most GIS applications are still dealing with 2D data, OGC standards explicitly support 3D geometries and the temporal dimension (essential for sensor data).

The standards and specifications developed by the OGC[4] thus make it possible to create complex and open information systems and services by promoting content and services accessible to all and usable by any type of web or local application.

8.3.3. *What standards for BIM – GIS convergence?*

BIM and GIS each develop a model and representation of the real world. BIM focuses on describing constructions over their entire lifecycle, from design, through construction, maintenance and possibly dismantling. It is based on a logic of design systems from which it borrows methods and tools (EXPRESS and STEP) to build its standards. The manipulated objects are a priori known from their design (except when building a BIM model of an old construction).

The world of GIS focuses on describing what is observable. In certain fields such as geology, knowledge is based a priori only on a few scattered observations and models are built progressively to gain precision as recognition progresses.

Even though the two fields naturally overlap, the construction processes for 3D BIM and GIS models are therefore fundamentally different. Thus, it seems illusory to implement them through a single standard covering all the data and processes.

BIM projects require access to various data sources providing information about their environment (cadastre, protected areas, geology, neighbors, flora and fauna, etc.). Most of this data are produced, managed and updated in a GIS environment, and often accessible through OGC standards. Rather than considering "remodeling" all of this data in an IFC grammar (BIM format), the approach that must be favored is to identify which data must really be integrated in a 100% BIM environment, and which data must "simply" be accessible for information or consultation from a BIM environment.

---

4 List of OGC standards are available at https://www.ogc.org/docs/is.

Conversely, a GIS project that requires access to information on constructions or infrastructure does not generally require access to all of the exhaustive information contained in the BIM project.

As with any interoperability approach, the process must begin with the identification of use cases that makes it possible to identify the necessary exchanges between BIM and GIS, and to deduce the standards to be used, extended or created (Fosu 2015).

However, there are mainly two OGC defined data exchange models and formats that have a strong overlap with IFCs: CityGML and InfraGML.

CityGML is a standardized open-data model and interexchange format for storing digital 3D models of cities and landscapes. It defines ways of describing most common 3D objects found in cities (e.g. buildings, roads, rivers, bridges, vegetation and street furniture) and their relationships. It also defines standards of different levels of detail for 3D objects, which allows the representation of objects for different applications and purposes, such as simulation, urban data mining, facilities management and thematic surveys. CityGML is built on GML. CityGML is widely used by software tools to represent the city-wide built entities in the GIS environment. CityGML also offers mechanisms to enrich the model with new types of data (Borrmann 2010; Laat and van Berlo 2011).

To address the needs of civil engineering, the OGC established a LandInfra SWG in 2013. This working group has taken elements from LandXML, an XML-based exchange format but not OGC standard, to develop a LandXML conceptual model, and InfraGML, as its implementation in GML. LandInfra covers the aspects of topography and the integration of surface and underground infrastructures, and especially the notion of alignment.

As the boundary between GIS and BIM standards is difficult to draw, there are natural tendencies in both domains to integrate elements already described in standards of the other domain. This is naturally the case between CityGML and LandInfra on the one hand and IFCs on the other. Studies have shown that the representations of an object in a model cannot be fully translated into the other formalism (Noardo et al. 2019) without losing part of their semantic or spatial richness (precision of coordinates in IFCs, for example).

8.3.4. *OGC – bSI Collaboration*

It is in this context that OGC and buildingSMART International (bSI) signed a partnership agreement in March 2014 with the primary objective of developing a common approach to the fundamental concept of alignment between InfraGML and IFC, through a common conceptual model. This partnership has since expanded with the creation of a joint working group called Integrated Digital Built Environment whose role is to coordinate joint work between the OGC and bSI. This work, which began on alignment, has notably extended to the representation of geotechnical data and, more broadly, geological observations related to structures.

Now it is clear that the shared objective is, rather than aiming for a single model for the representation of the natural and built environments (as described in Figure 8.5), to define how to ensure the best communication between the different domains, starting from use cases and processes (Breunig et al. 2020).

The approach currently favored is that of the semantic web (Usmani et al. 2020), which allows objects (entities) to be shared with all their semantic richness, and data exchanges being facilitated by the development of a common ontology (or more simply a common conceptual model).

These joint works find their materialization in the development of digital twins at the scale of an infrastructure or even a city (Nativi et al. 2020).

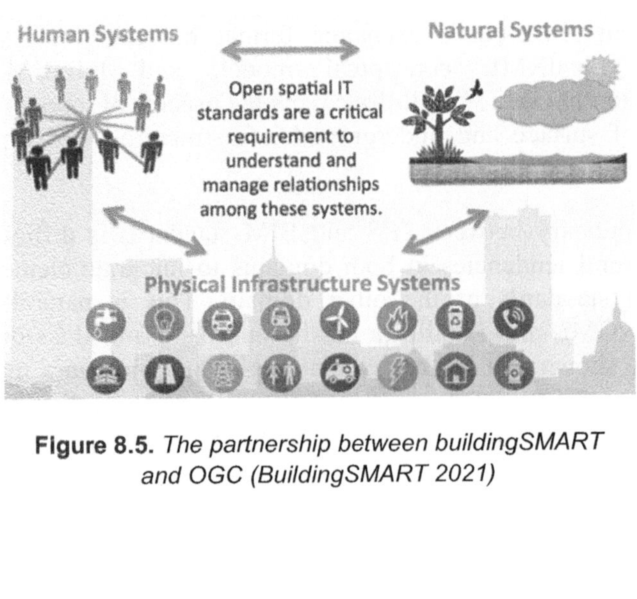

Figure 8.5. *The partnership between buildingSMART and OGC (BuildingSMART 2021)*

8.4. BIM and GIS interoperability

8.4.1. *Digital continuity*

It is an important parameter of the interface between BIM and GIS. As mentioned earlier, data are recognized as the essential component of both GIS and BIM. Today, it is fed by more and more diverse sources and in exponential quantities (big data, massive data). For example, the acquisition of data (topography, infrastructures, etc.) by a LIDAR type technology (laser scanner, drone, etc.) produces a quantity of points (point clouds) that usually exceeds millions, or even one billion. The ease in using these new data acquisition and sensing technologies makes it possible to use them at various scales, from bridge to tunnel, building to road, neighborhood to city, etc., to develop a BIM-type digital model as well as a GIS.

The acquisition of data from multiple sensors and Internet of Things (IoT) linked to a building or a territory (energy consumption, air quality, weather, etc.) concerns both GIS and BIM. IoT is a computing concept for devices that relate to each other over the Internet to communicate and are responsible for sending and receiving data. IoT devices provide real-time data that present a powerful paradigm for the Smart City application to be integrated with BIM and GIS systems. IoT enables technologies including sensing, communication, identification and recognition technologies with the ability to share information across platforms through a unified framework. Integration of IoT technology with infrastructure information is complementary and enables innovative applications.

The potential integration of BIM and GIS with IoT devices enables new possibilities for better decision-making of spatial problems throughout the lifecycle of urban and environmental processes. It results in a detailed 3D city model fed with building data, city infrastructure and information sampled from sensors as a smart city for spatial analysis and integrated reasoning at multiple levels of detail. IoT devices compliment both BIM and GIS. However, BIM and IoT integration still appears to be at an early stage, even with integration methods involving relational databases using new schema, new query languages, semantic web technology and a hybrid approach (Tang et al. 2019; Zhu et al. 2020). Furthermore, integrating GIS, BIM and IoT leads to smart management (Park et al. 2018) including AR and VR technologies (Carneiro et al. 2019). Despite great interest to use IoT and BIM together in the construction industry and facility management, the question of integrating GIS information with BIM still remains open and challenging. BIM-GIS is an integral source for infrastructure applications, and studies with IoT associated to such integration methodologies are limited.

Figure 8.6 shows this digital continuity and the integration of a variety of new data (both in a GIS and in a digital model).

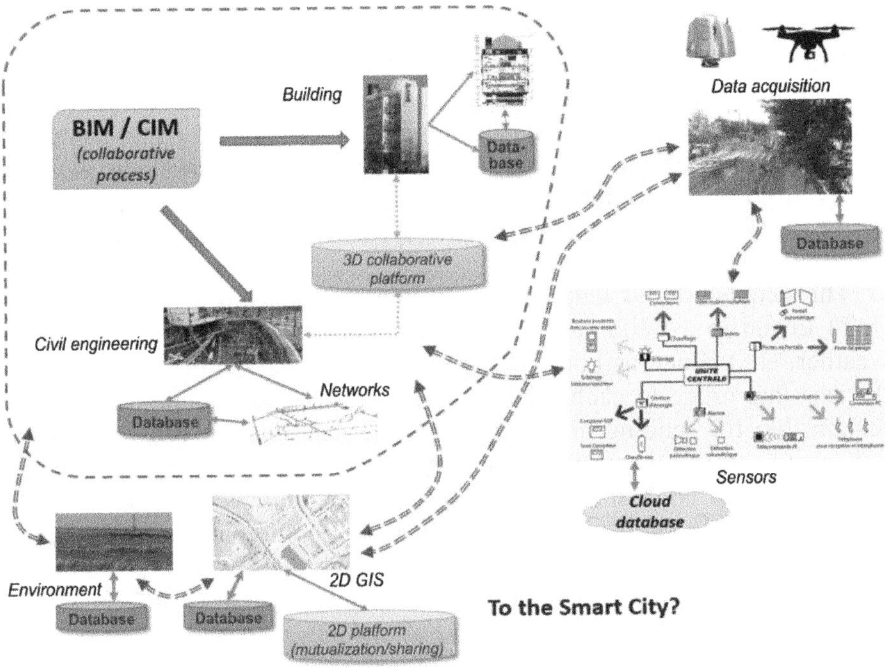

Figure 8.6. *Digital continuity and new data integration in GIS and BIM*

8.4.2. *Exchange formats versus interoperability*

An exchange format is a form of interoperability. It allows competing software (different editors) to exchange data (GIS, 3D modeling of linear infrastructures, digital models of buildings, etc.) using a common external format. It is a result of a consensus of discussions and IT developments, most often based on standardization/standards. The CityGML for 3D GIS and the IFC for BIM (buildingSMART) are particularly useful exchange formats, in their own domains. They allow users to exchange structured data with associated information. The limitation of the exchange formats currently lies in the fact that exporting data from one proprietary format (sender) to another proprietary format (receiver), via an intermediate exchange format, produces duplications or loss of data. If that is not a problem (except possibly for storage or loss of useful information), updating it becomes complicated, because the source file and its duplications must be systematically processed (Figure 8.7).

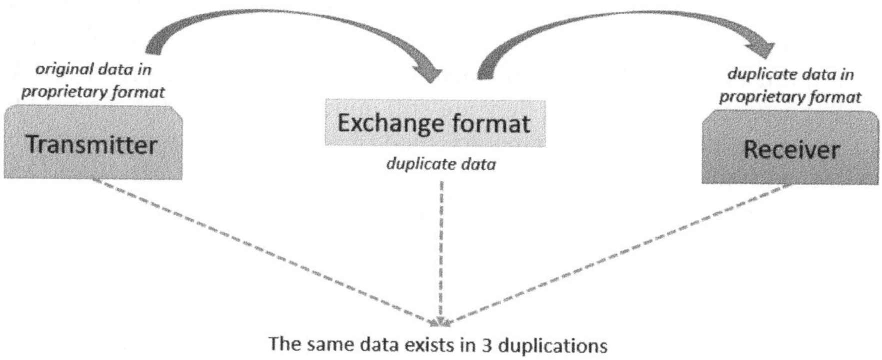

Figure 8.7. *Data duplication scheme*

Full interoperability, on the contrary, eliminates the intermediate step of the data exchange format because everything then takes place in a particular RDBMS. Here, data are stored in a single format and are accessible by the software of the different editors, without any transformation of original data. It is no longer software and exchange formats that condition access to the data; it's the RDBMS (spatial server type in GIS) that prevails. Therefore, is the exchange format opposable to interoperability? No, of course not! In fact, there are several levels of interoperability:

– Initial level: Import/export functionalities between software, which allow data to be exchanged without any guarantee that there is no loss of information (DXF for CAD/CAM or SHP for GIS).

– Level 1: Interoperability between two competing software programs, when the publishers of these programs have decided to interface together some of their developments, so that their users can exchange data easily without worrying about a question of format.

– Level 2: Exchange format recognized by a business community (GIS, building, civil engineering) as an intermediate key, and giving rise to standardization (CityGML, IFC, InfraGML, etc.).

– Level 3: Access to data without format transformations and without duplication (spatial server for GIS, collaborative platforms, etc.).

8.4.3. *The new collaborative tools*

The software and business applications commonly using GIS and BIM are nowadays mature and their use by professionals is long-established – with a limit, however, because if open-source tools are widespread in GIS (QGIS, PostGIS, GvSIG, etc.), they are still quite limited in BIM (FreeCAD, etc.). On the other hand, participatory mode has been used for a long time in GIS, and the collaborative mode is less used where all users can interact/revise/add the input to the systems. It is the latter that prevails in BIM, with the use of the digital model. There is, however, a striking and specific element of BIM that is developing today: the use of collaborative platforms, which allow continuous exchanges and effective collaboration in the building and infrastructure trades (in GIS, they are rather platforms for data sharing/mutualization). Another new phenomenon is that these platforms are no longer simply the prerogative of digital modeling software publishers; other players are successfully positioning themselves in this niche. Interoperability within these platforms is not yet optimal, but they can manage different exchange formats, by upgrading their level of interoperability. In the near future, it is possible that collaboration will not only take place on one platform, but also across several, interconnected platforms according to their specificities. We will then talk about both collaborative platforms and platform collaboration.

8.4.4. *The evolution of practices and skills*

Integration of BIM and GIS domains with sensing is emerging as an important area of research but the two offer different visions for the interpretation of the 3D model. Recently, a series of studies were conducted as a critical and state-of-the-art review on BIM-GIS integration methods (Deng et al. 2016; Liu et al. 2017). By complimenting their strength and weakness of most relevant integration models, potentialities of other models are classified based on parameters elected for integration. Efforts toward data interoperability with a focus on BIM and GIS integration were made using prominent data exchange formats of two domains: Industry Foundation Classes (IFC) and CityGML (Hor et al. 2016; Usmani et al. 2020). However, studies are still being conducted to exchange information between BIM and GIS as they both need information from each other to enrich their work.

Previous studies have shown increasing consideration in integration technologies of various approaches and models for BIM-GIS integration to come up with efficient means of resolving integration problems (Liu et al. 2017; Breunig et al. 2020; Noardo et al. 2020). For the integration pattern of BIM and GIS, studies were proposed using unidirectional approaches (extract BIM data to GIS or GIS to BIM), some proposed new tools, frameworks, extensions, and ontologies, while few considered bidirectional methodology as Unified Building Model (El-Mekawy and Östman 2010; El-Mekawy et al. 2012). However, the extraction and simplification of data during the process of integration from one domain to another results in information loss (Hor et al. 2018; Noardo et al. 2020). Interoperability of IFC and CityGML is mapping between key schemas as they have different elementary development purposes, concepts and structures. BIM-GIS integration has undergone various perspectives among schema mapping (El-Mekawy and Östman 2010; Deng et al. 2016), ontological modeling (Hor et al. 2016), integrated web services (Karan and Irizarry 2015), data transformations and schema extensions (Noardo et al. 2020). A state-of-the-art review on GIS and BIM integration by Liu et al. (2017) compared the selection of method against four parameters: effort, extensibility, effectiveness and flexibility (EEEF). The information loss, incompatibility of available software and data formats, and limitations in use-case-specific frameworks are common shortcoming of existing approaches. However, Semantic Web technologies have been identified as promising compared to other methods presented in Table 8.2.

Integration methods	Effectiveness	Extensibility	Effort	Flexibility
New standards and models	Case by case	Case by case	Case by case	Case by case
Conversion, translation and extension (manual)	Medium	High	High	Medium
Conversion, translation and extension (semi-automatic)	Medium	Medium	Medium	Medium
Semantic web technologies	High	High	High	Medium
Services-based methods	High	Low	High	Low
Application focused methods	Case by case	Low	Low	Low

Table 8.2. *Comparison of BIM and GIS data integration methods and EEEF by Liu et al. (2017)*

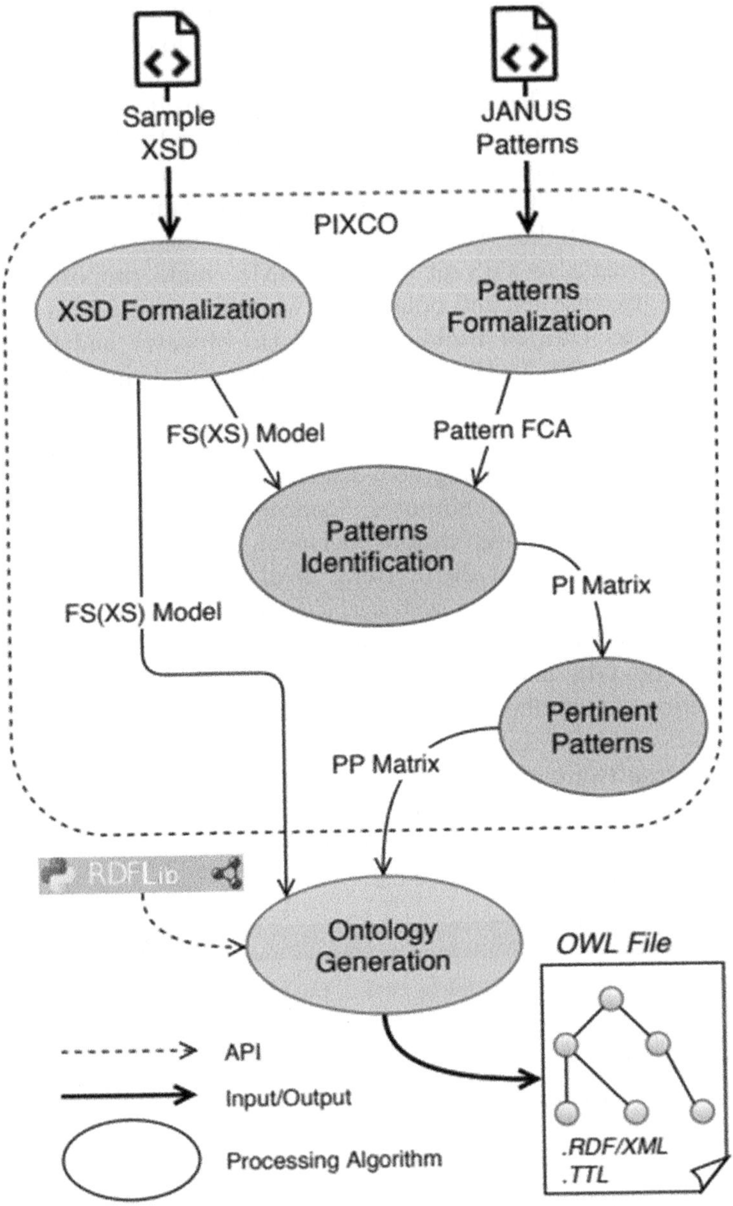

Figure 8.8. *Semi-automatic ontology generation framework extracted from Usmani et al. (2020)*

Semantic Web technology establishes information exchange across independent and fundamentally incompatible data formats. Semantic Web and Linked Data have been investigated by researchers as complementary for technologies in the existing AEC industry (Pauwels and Terkaj 2016). Integration methods proposed by Hor et al. (2016) and Karan et al. (2016) enable enhanced data exchange and integration between BIM and GIS at the semantic level. Integration of heterogeneous data by ontology modeling is an effective approach in semantic web, consequently comprising ontologies for BIM and GIS standardized formats, IFC and CityGML. Past studies have been conducted to develop and standardize the ontology models of IFC (Pauwels and Terkaj 2016) and CityGML (Métral et al. 2013; Zalamea et al. 2013; Wang and Issa 2020). Pauwels and Terkaj's approach (2016) provides extensive framework for EXPRESS based IFC to OWL, and the CityGML ontology generated by Métral et al. (2013) needs human intervention. It is being highlighted by Usmani et al. (2020) that XML-based common format can be adapted among BIM and GIS, that is, ifcXML and CityGML, to provide solutions comparable within the same context. In the same study, Usmani et al. (2020) introduced a semi-automatic approach to generate BIM ontology (O_{BIM}) and GIS ontology (O_{GIS}) by extending Janus (Bedini et al. 2011) and PIXCO (pattern identification for XSD conversion to OWL) (Hacherouf et al. 2019) as presented in Figure 8.8.

Linking BIM and GIS ontologies with other sensor domain ontologies is an effective approach for heterogenous data integration. IoT, on the other hand, with divergent sensor data sources, are kept toward further research to be mapped as IoT ontology (O_{IoT}) with O_{BIM} and O_{GIS} as introduced by Usmani et al. (2020). This leads to an extended unified ontology that serves complete interoperability toward BIM, GIS and IoT semantic data modeling involving semantic and structural alignment techniques. For instance, machine learning approaches such as semantic-based Word2vec (Mikolov et al. 2013) and structure-based Node2vec (Grover and Leskovec 2016) algorithms are suggested by the literature for ontology mapping relations (e.g. one-to-many, many-to-one or many-to-many). Indeed, OWL representation of ontology assists reasoning and inference. However, resource descriptive framework (RDF) graphs are the core of the semantic web and annotate to linked data. RDF graphs are next steps to generate IFC and CityGML datasets using respective generated ontologies. The similarity of cross-domain ontology is to bridge the gap between BIM and GIS providing an interlinked RDF graph, a common data model in the semantic web. In addition, IoT data as linked data can potentially be interlinked with this common data model through subject-predicate-object data model. RDF graphs produced for such semantically rich geospatial and IoT information are considerably sizable and likely require optimizations for querying, storing, and managing operations. To address these, cluster-based and community optimization techniques can be applied for manipulating RDF graphs with exhaustive information. As a complete process, this leads to the interoperability of fundamentally distinct domains such as BIM, GIS and IoT data. The IoT information integration with BIM and GIS data is reaching extreme interest for smart city applications such as Digital Twin development, accordingly, evolving toward Smart Data, a machine-readable information. Automation is key in the current situation toward knowledge discovery, effective planning and informed decision-making, leading the urban innovation in lifestyle, environment, and mobility with smart data.

8.5. Conclusion and perspectives

GIS and BIM are rooted in different communities, and now there is a great potential for integrating these two data sources in urban analytics, land management, infrastructure and building monitoring. It is significant that no BIM and GIS communities aim to replace each other; rather, integrating these two data sources would be beneficial for both parties. Both communities would come to exploit BIM data within a broader context (i.e. city level) with the help of GIS data and tools from building, construction sites, infrastructure and urban analytics. On the technical level, each area brings experiences and strengths such as:

– GIS provides standards to develop service-oriented architecture, database management systems, spatial analysis techniques, and visualization platforms.

– BIM provides a rich semantic description of building components, building process and lifecycle, the dynamics between various components of BIM, and collaboration among different stakeholders through the lifecycle of the building (e.g. architecture, engineers, facility management, to name a few).

Indeed, the exchange and conversion of formats between BIM and GIS received attention for more than decades. With the advancement of semantics web technology and ontology generation, a promising avenue has risen in exchanging a specific domain for another one by relying on their unique data storage and access to specific views according to the end-users' needs. In addition, integrating BIM and GIS data with IoT data will open the opportunity to realize the Digital Twins idea at different scales from building to the city. Likewise, all digital transformations, integrating BIM and GIS, will have a direct impact on the professions and end-users by providing decompartmentalization of disciplines, allowing the development of new applications, and better and informed responses to existing challenges. In terms of technology enhancement, developments of BIM and GIS have been increasingly driven by application software developers. This brings extensive attention to explore the interoperability in multiple stages from data providers, data management systems, data analytics and visualization, sharing and collaboration platforms and, finally, digital twin development. However, we are still in early stages, but we can hope for the arrival of new professionals with advanced digital transformation skills and analytical mindsets. This may lead to less dependence of BIM and GIS communities on software suppliers; instead, the focus can be on quality of data, consistency and relationships with other existing and historical data.

8.6. References

Bedini, I., Matheus, C., Patel-Schneider, P.F., Boran, A., Nguyen, B. (2011). Transforming XML schema to OWL using patterns. In *Proceedings – 5th IEEE International Conference on Semantic Computing, ICSC 2011*. doi:10.1109/ICSC.2011.77.

Bordin, P. and Gimenez, L. (2017). SIG & BIM : quelle histoire ! *Géomatique expert*, 115, 38–43.

Borrmann, A. (2010). From GIS to BIM and back again – A spatial query language for 3D building models and 3D city models. *International Archives of the Photogrammetry, Remote Sensing and Spatial Information Sciences*, 38.

Breunig, M., Bradley, P.E., Jahn, M., Kuper, P., Mazroob, N., Rösch, N., Al-Doori, M., Stefanakis, E., Jadidi, M. (2020). Geospatial data management research: Progress and future directions. *ISPRS International Journal of Geo-Information*, 9(2), 95.

buildingSMART International (2021). buildingSMART OGC FutureCities Collaboration [Online]. Available at: https://www.buildingSMART.org/buildingSMART-ogc-futurecities-collaboration/ [Accessed 22 May 2021].

Carneiro, J., Rossetti, R., Silva, D.C., Oliveira, E.C. (2019). BIM, GIS, IoT, and AR/VR integration for smart maintenance and management of road networks: A review. In *IEEE International Smart Cities Conference, ISC2 2018*. doi:10.1109/ISC2.2018.8656978.

De la Beaujardiere, J. (ed.) (2002). OpenGIS® Web Map Server Implementation Specification Version 1.1.1. Open Geospatial Consortium Inc.

Deng, Y., Cheng, J.C., Anumba, C. (2016). Mapping between BIM and 3D GIS in different levels of detail using schema mediation and instance comparison. *Automation in Construction*, 67, 1–21.

El-Mekawy, M. and Östman, A. (2010). Semantic mapping: An ontology engineering method for integrating building models in IFC and CITYGML. *Proceedings of the 3rd ISDE Digital Earth Summit*. doi:10.1021/bm049735c.

El-Mekawy, M., Östman, A., Hijazi, I. (2012). An evaluation of IFCCityGML unidirectional conversion. *International Journal of Advanced Computer Science and Applications*, 3(5). doi:10.14569/IJACSA.2012.030525.

Fosu, R., Suprabhas, K., Rathore, Z., Cory, C. (2015). Integration of Building Information Modeling (BIM) and Geographic Information Systems (GIS) – A literature review and future needs. In *Proceedings of the 32nd CIB W78 Conference 2015*. Eindhoven.

Goodchild, M.F. (2010). Twenty years of progress: GIScience in 2010. *Journal of Spatial Information Science*, 1. doi:10.5311/JOSIS.2010.1.2.

Goodchild, M.F., Yuan, M., Cova, T.J. (2007). Towards a general theory of geographic representation in GIS. *International Journal of Geographical Information Science*, 21(3), 239–60.

Grover, A. and Leskovec, J. (2016). Node2vec: Scalable feature learning for networks. In *Proceedings of the 22nd ACM SIGKDD International Conference on Knowledge Discovery and Data Mining, KDD*. doi:10.1145/2939672.2939 754.

Hacherouf, M., Nait-Bahloul, S., Cruz, C. (2019). Transforming XML schemas into OWL ontologies using formal concept analysis. *Software and Systems Modeling*, 18(3). doi:10.1007/s10270-017-0651-4.

Hor, A.H., Jadidi, A., Sohn, G. (2016). BIM-GIS integrated geospatial information model using semantic web and rdf graphs. *ISPRS Annals of the Photogrammetry, Remote Sensing and Spatial Information Sciences*, 3, 73–79.

Hor, A.E., Sohn, G., Claudio P., Jadidi, M., Afnan, A. (2018). A semantic graph database for BIM-GIS integrated information model for an intelligent urban mobility web application. *ISPRS Annals of the Photogrammetry, Remote Sensing and Spatial Information Sciences*, 4(4), 89–96.

Irizarry, J. and Karan, E.P. (2012). Optimizing location of tower cranes on construction sites through GIS and BIM integration. *Journal of Information Technology in Construction*, 17, 351–366.

Karan, E.P., Irizarry, J., Haymaker, J. (2016). BIM and GIS integration and interoperability based on Semantic Web technology. *Journal of Computing in Civil Engineering*, 30(3). doi:10.1061/(ASCE)CP.1943-5487.0000519.

Laat, R. and van Berlo, L. (2011). Integration of BIM and GIS: The development of the CityGML GeoBIM extension. In *Advances in 3D Geo-Information Sciences*, Kolbe, T.H., König, G., Nagel, C. (eds). Springer, Berlin, Heidelberg.

Liu, X., Wang, X., Wright, G., Cheng, J., Li, X., Liu, R. (2017). A state-of-the-art review on the integration of Building Information Modeling (BIM) and Geographic Information System (GIS). *ISPRS International Journal of Geo-Information*, 6(53). doi:10.3390/ijgi6020053.

Métral, C., Ghoula, N., Silva, V., Falquet, G. (2013). A repository of information visualization techniques to support the design of 3D virtual city models. In *8th 3D GeoInfo Conference*, 2(2). doi:10.5194/isprsannals-II-2-W1-247-2013.

Mikolov, T., Sutskever, I., Chen, K., Corrado, G., Dean, J.R. (2013). Distributed representations of words and phrases and their compositionality. In *Advances in Neural*, Burges, C.J.C., Bottou, L., Welling, M., Ghahramani, Z., Weinberger, K.Q. (eds). Curran Associates Inc., Red Hook, NY.

Nativi, S., Delipetrev, B., Craglia, M. (2020). Destination Earth survey on "Digital Twins" technologies and activities. *The Green Deal Area*. doi:10.2760/430025.

Nebert, D. (2001). *Developing Spatial Data Infrastructures. The SDI Cookbook.* Global Spatial Data Infrastructure.

Newell, A. and Simon, H.A. (1972). *Human Problem Solving.* Englewood Cliffs, Prentice Hall, Hoboken, NJ.

Noardo, F., Harrie, L., Ohori, K., Biljecki, F., Ellul, C., Eriksson, H., Guler, D., Hintz, D., Jadidi, M., Pla, M. et al. (2020). Tools for BIM-GIS integration (IFC georeferencing and conversions): Results from the GeoBIM Benchmark 2019. *ISPRS Int. J. Geo-Inf.*, 9(502).

Pauwels, P. and Terkaj, W. (2015). EXPRESS to OWL for construction industry: Towards a recommendable and usable ifcOWL ontology. *Automation in Construction*, 63, 100–133. doi:10.1016/j.autcon.2015.12.003.

Reichardt, M. and Robida, F. (2019). Why standards matter – The objectives and roadmap of the International Open Geospatial Consortium (OGC). *Annales des Mines – Responsabilité et environnement*, 94(2), 25–29. doi:10.3917/re1.094.0025.

Tang, S., Shelden, D.R., Eastman, C.M., Pishdad-Bozorgi, P., Gao, X. (2019). A review of building information modeling (BIM) and the internet of things (IoT) devices integration: Present status and future trends. *Automation in Construction*, 101, 127–139. doi:10.1016/j.autcon.2019.01.020.

Usmani, A.U., Jadidi, M., Sohn, G. (2020). Automatic ontology generation of BIM and GIS data. *ISPRS – International Archives of the Photogrammetry, Remote Sensing and Spatial Information Sciences*.

Wang, N. and Issa, R.R.A. (2020). Ontology-based integration of BIM and GIS for indoor routing. doi:10.1061/9780784482865.107.

Zalamea, O., Orshoven, J., Steenberghen, T. (2013). From a CityGML to an ontology-based approach to support preventive conservation of built cultural heritage. In *AGILE International Conference on Geographic Information Science*, Helsinki.

Zhu, J., Wright, G., Wang, J., Wang, X. (2018). A critical review of the integration of geographic information system and building information modelling at the data level. *ISPRS International Journal of Geo-Information*, 7(2). doi:10.3390/ijgi7020066.

Glossary

4D model, section 6.3.1: A 4D model is a 3D model linked to the construction schedule (site planning and production planning).

5D model, section 6.3.1: A 5D model is a model linked to costs and resources (resource planning).

5S method, section 6.2.2: A management method that is part of the quality approach. Developed within the framework of the Toyota production system, it takes its name from the first letter of each of five operations constituting as many simple words of order or principles: *Seiri* (tidy), *Seiton* (order), *Seiso* (cleaning), *Sei-ketsu* (clean) and *Shitsuke* (education).

A3 problem-solving method, section 6.2.2: The method requires prioritizing and formulating the problem and its solution succinctly on an A3 sheet of paper. The A3 method embraces the Plan-Do-Check-Act (PDCA) cycle.

Abstract design, sections 4.1 and 4.3: Abstract design is based on the notion of model and modeler. The circulation of generic and prescriptive abstract models between teams, the distribution of work by codified processes, allows for the generation of a generic model. This model is able to describe both the artifact, called the "system of interest" and its design process, called the "enabling system". It is necessary to ensure the continuity of interpretation between numerous concepts and models of thought: requirements management, the entry point for satisfying the needs of the client, systems engineering, conceptual data models specific to the various trades, modeling formalisms and exchange formats. It is therefore a question of moving from data management via modeling to information management via standardization in order to move toward knowledge management (via semantic interoperability), all implemented through a continuous digital process (digital twin).

Adaptive reuse potential (ARP), section 7.3.4.2: This method assists in the transformation of the traditional decision-making processes of property stakeholders toward more sustainable practices, strategies and outcomes, by providing a means by which the industry can identify and rank existing buildings that have high potential for adaptive reuse. This method was developed by Langston et al. (2008).

AdaptSTAR, section 7.3.4.2: AdaptSTAR offers holistic and unified design criteria suitable for assessing the adaptive reuse potential (ARP) of future buildings. The findings show that criteria can be identified and weighted according to physical, economic, functional, technological, social, legal and political categories to calculate an adaptive reuse star rating (Conejos et al. 2013).

Agile approach *(Agile Manifesto)*, section 2.7: A set of concepts and techniques for collaborative design, promoting continuous data sharing and regular interactions among stakeholders. Coming from agile manufacturing and software project management, the agile approach is nowadays used in numerous industries, particularly the construction sector.

Analytical or as-built model, section 1.5: It is still called dynamic since it allows for simulations.

Architecture engineering construction (AEC) industry, section 1.2: Three types of organizations are interacting to build BIM in the construction industry, that is, project owners, construction actors (designers and builders) and editors. In the construction industry, the project-based approach is long standing, very well established and very structuring. This facilitates the gradual adoption of technological change and the gradual adjustment of the company's resources (Christensen and Overdorf 2000). The companies cooperate on building sites, each allocated a set of tasks. On the construction site, they are obliged to cooperate and exchange data, while keeping their individual commitments, which also provides a strong incentive for a structured exchange of data, and thus for the adoption of BIM.

Artificial intelligence (AI), section 6.3.1: "Artificial intelligence is a capability of a functional unit to perform functions that are generally associated with human intelligence such as reasoning and learning. Note 1 to entry: artificial intelligence; AI: term, abbreviation and definition standardized by ISO/IEC" (ISO/IEC 2382-28:1995).

Bill of quantities (BOQ), section 6.3.1: It is a document used in construction tenders in which materials, parts and labor (and their costs) are detailed.

BIM, section I.1.2: The National BIM Standard – United States® (NBIMS) has ascribed three meanings to the acronym BIM: building information modeling, model or management (BIM3) (NBIMS 2015). In this book, we have chosen building information modeling as the sole meaning, and we refer to the definition outlined in ISO 19650: "a shared digital representation of a built asset (3.2.8) to facilitate design, construction and operation processes to form a reliable basis for decisions. Note 1 to entry: Built assets include, but are not limited to, buildings, bridges, roads, process plants. ISO 29481-1 (3.2, modified, 2016) – The word 'object' has been replaced with 'asset'; the words 'including buildings, bridges, roads, process plants, etc.' have been removed; original Note 1 to entry has been replaced with a new one" (2018).

BIM-DAS, section 7.3.4.2: "The BIM-based Deconstructability Assessment System (BIM-DAS) provides an objective and measurable system for building deconstructability during the design stage. This scoring system forms a basis for comparative analysis building models to choose the option with the least end of life impact on the environment". It was developed by Akinade et al. (2015).

BIM-to-FM, section 7.2.3: "Regarding interoperability of BIM with other software, latest developments focus on BIM-to-FM information interoperability in standardization (e.g., CoBie standard) or research (see e.g., Chen et al. 2018)".

BIM adoption (conceptual structures), section 1.4: Succar and Kassem (2015), studying the conceptual structures of BIM adoption, distinguish between BIM implementation – the successful adoption of BIM tools and a BIM approach in an organization – and BIM diffusion – the rate of use of the tools and data exchanges across the market. BIM implementation is seen as occurring in three phases: readiness to adopt, capability to perform and performance maturity.

BIM approach, section 1.2.4: A "BIM approach" is more than just a new generation of software; adoption processes must be implemented in addition to the deployment from an IT perspective of new tools and innovations. The precursor aspect of the innovation changes the roles in its adoption process (Ben Mahmoud-Jouini and Charue-Duboc 2014).

BIM as a technological disruption, section 1.1.3: BIM as a technological disruption initially appeared to be "imposed" by the evolution of digital technology. The product moves from the delivery of a physical work with large sets of interoperable and organized data onto one where services are grafted. The BIM is organized above all around the data, and it concerns the entire construction project, and therefore all the trades and companies that the project brings together. This implies cooperation and data sharing between all the companies involved in the project and recomposes the creation of added value, the roles and the interrelations of all the stakeholders.

BIM execution plan (BEP), sections 4.2.2 and 1.5: It is a plan that explains how the information management aspect of the consultation will be carried out by the execution team. The BEP is constrained by the system "to be made" and the system "for making it" (ISO 19650-2, 2018). The BEP has as inputs the product requirements to identify the uses of BIM, and it includes the description of process requirements and modeling requirements.

BIM methodology, section 4.1: BIM as a methodology for projects brings together the consequences of the paradigm shift (Tolmer 2016). The new needs in modeling and formalisms, the system vision of a project, as well as the management of exigences, are some of these consequences. It requires a global thinking on project information, through data models and improvements in interoperability.

BIM maturity, section 1.5: The first edition of the NBIMS standard (2007) enabled evaluating the maturity of companies in relation to BIM by proposing a structure through maturity levels. Since then, several BIM maturity assessment tools have been proposed, the precursors being, in particular: the BIM Capability MM (BIM CMM) (NBIMS 2007), the iBIM maturity model (Bew et al. 2008) and the BIM Maturity Matrix (Succar 2009; Succar et al. 2012). These different models are the subject of numerous comparative studies (Giel and Issa 2014; NBIMS 2015; Wu et al.

2017; Ferraz et al. 2020). The UK model – known as the iBIM or BIM Wedge model (Bew et al. 2008) – exists in several versions, including the one published in the UK Government Construction Client Group (GCCG) report in 2011.

BIM maturity levels, section 1.5: The iBIM model identifies specific capability levels covering technology, standards, guides, classifications and delivery, which show the progress of BIM adoption by the organization. Level 0: 2D CAD, unstructured, typically paper-based. Level 1: A mix of 2D CAD and 3D digital mockup. The data are structured and the process includes a collaboration tool. Level 2: 3D environment managed in separate BIM tools, specific to each collaborator (architect, design offices, builders), with structured data attached and possible exchanges of digital models. Level 2 allows the use of BIM 4D and 5D (BIM 4D model to which we added the data related to the cost). Level 3: this is the highest maturity level. It corresponds to iBIM or "integrated BIM". The integration of processes and data is complete. It is close to a concurrent engineering process. A single model is stored on a centralized server, accessible by all participants throughout the lifespan of the work via formats such as IFC, CityGML, BCF and methodological tools and structuring of information and methods. The maturity levels are now understood by the exchange of standards and by the access to database, as a heterogeneous environment. The maturity levels must be seen in relation to the standards. BS 1192, superseded by the ISO 19650 (2018) series of standards, is the first element that defines Level 3. It corresponds to the exchange of 3D data and implies this exchange framed by standards.

BIM specialists, section 1.2.2: Several proposals have been made for naming and defining the competencies of BIM specialists. For example, Gu and London propose the four roles of BIM specialists: BIM manager, information manager, BIM coordinator and BIM technician. Based on project-centric skills and organization-centric skills, and by cross-referencing skill sets and elementary skills in the project, they construct four profiles.

Building description systems (BDS), section I.1.1: A concept introduced by Eastman et al. (1975), covering a prototype to develop a general building description system, since recognized as the precursor of BIM tools, in several steps to the BIM process (Latiffi et al. 2014): GLIDE: Graphical

Language for Interactive Design (Eastman and Henrion 1977) – BPM: Building Product Model (Bjork 1989) – GBM: Generic Building Model (Eastman and Siabiris 1995).

C&D, section 7.1: Construction and deconstruction sector.

Capability maturity model (CMM), section 1.5: Model proposed by the Software Engineering Institute (Paulk et al. 1993). CMM takes a different approach from the maturity matrix, by looking at key performance indicators (KPIs), which should be achieved as the maturity level increases. Fraser et al. (2002) identify a third CMM based on a questionnaire and a Likert-type (1932) rating scale to assess the integration maturity of a process in a quantified way.

Circular economy, BIM maturity levels, section 1.5: "If a building component/product or a whole part of a building is reused, then it re-enters the circular economy in a new use phase. The functional reuse of whole building components and products can be performed after separation, decommissioning and further processing steps. It can take place at the same site (e.g., in a replacement building) or at a new site in the installation of a new building or infrastructure. Recycling defines the reuse of building materials, components and products after the destruction of their structure and post-processing to a new component or product made from only or partly secondary raw material (building as a secondary mine or material bank)".

CityGML, section 8.3.3: It is a standardized open data model and interexchange format for storing 3D digital models of cities and landscapes. It defines ways to describe the most common 3D objects found in cities (e.g. buildings, roads, rivers, bridges, vegetation, water body, and street furniture) and their relationships. It also defines standards for different levels of details for 3D objects, which allows the representation of objects for different applications and purposes, such as simulation, urban data mining and visual analytics, facility management and thematic surveys. CityGML is built on GML. CityGML is widely used by software tools to represent the city-wide built entities in geographic information systems (GIS) environment. CityGML also provides mechanisms to enrich the model with new data types (Borrmann 2010; Laat and van Berlo 2011).

Closed-Loop PLM, section 2.3.2: A Closed-Loop PLM system, or Closed-Loop Lifecycle Management, allows all stakeholders who play a role during the lifecycle of a product (project managers, designers, manufacturers, maintenance operators, recyclers, etc.) to track, manage and control product information at any phase of the lifecycle (design, manufacturing, Middle of Life and End of Life), at any time, and from any location worldwide (Kiritsis 2011).

Cloud computing, section 6.3.1: "Cloud computing is a model for enabling ubiquitous, convenient, on-demand network access to a shared pool of configurable computing resources (e.g., networks, servers, storage, applications, and services) that can be rapidly provisioned and released with minimal management effort or service provider interaction".

Common data environment (CDE), sections 1.4 and 1.5: It is an agreed-upon source of information about a given project or asset, used to collect, manage and disseminate each piece of information through a managed process (BS 1192:2007). The adoption of a BIM approach, by a company, requires reinventing the processes implemented in projects. It also depends on the common data environment that can be adopted at the project level itself, by the inter-company teams.

Computation independent model (CIM), section 5.2.1: The computation independent model is concerned with the context and requirements (prerequisites) of the system, without taking into account its structure or processes. The UML language (Booch et al. 1998) is used to define this type of model (ISO/IEC 19505, 2012).

Concurrent engineering, section 2.2.2: It is "an engineering method that consists of involving all the actors of a project, from the beginning of the project, in the understanding of the objectives sought and of all the activities that will have to be carried out" (author's translation). It became a standard practice in the manufacturing industry in the mid-1990s (Sohlenius 1992).

Continuous improvement, section 6.3.3: Continuous improvement (CI), which is at the heart of Lean (Womack et al. 1990), is essential to the success of any FM organization (Beck et al. 2016). CI is an approach by which (a) small incremental improvement steps are taken to improve performance (Slack et al. 2010) and (b) waste in all processes of an organization are identified, reduced and eliminated (Bessant et al. 2001).

Cooperation, section 1.2.3: The BIM approach implies, through cooperative processes, integrating the understanding of the partner's objectives, which implies a deep understanding of the other's specialty or trade and is an essential component of cooperation (Zacchary and Robertson 1990). This combination of heterogeneous knowledge from different worlds (Yoo et al. 2012) profoundly renews all modes of collective work and cooperation in the project.

Core product model (CPM), section 2.4.1: It proposes a generic product model based on artifacts (product components) aggregating the three views of function, form and behavior (Fenves et al. 2008). It was initially developed at NIST (National Institute of Standards and Technology).

Data, sections 4.2 and 5.1: "Data are observations or items obtained from measurements of variables" (ISO 772, 2011). Data have attributes of propriety, accuracy and completeness. They follow a formal syntax and use concepts defined in dictionaries (ISO/DIS 8000-1, 2011).

Data dictionaries, section 5.1: Data dictionaries are informal, since they contain only terms and their definitions. The interpretation of these terms (their semantics) is the responsibility of the user and cannot be done by the machine. This is also the case for glossaries and term hierarchies.

Data model languages, section 5.1: Data model languages (e.g. UML or Unified Modeling Language) are more formal than those used to define a dictionary. The rules for composing elements (syntax) are more constrained. However, the meaning of the data is not formally specified, and the data models represent (graphically) knowledge (for this reason they are said to be meaningful). However, it is important to note that an algorithm cannot reason with this knowledge, and a domain expert is required for interpreting it.

Database law, section 3.3.2: A BIM collaborative database contains large volumes of information generated by many contributors. Database law should now apply to BIM as a composite work, owned by the author who made the integration, subject to the copyright of the pre-existing work.

Digital continuity, sections 1.2.2 and 8.4.1: Digital continuity, which is not specific to construction, imposes the use of data by a chain of actors and repositions the role of each. Digital continuity forces us to reconsider and break down the channels of activities and knowledge. It is a revolution in the way we deliver information, not just in terms of the tools. Digital continuity is an important parameter of the interface between BIM and GIS. Data are recognized as the essential component of GIS and BIM. Today, it is fed by more and more diverse sources and in exponential quantities (Big Data, massive data). For example, the acquisition of data (topography, infrastructure, etc.) by a LIDAR type technology (laser scanner, drone, etc.) produces an enormous point cloud that generally exceeds millions, even billions of points. Facilitating the use of these new data acquisition and detection technologies allows them to be applied at different scales; from bridges to tunnels, from buildings to roads, from neighborhoods to the cities, etc., in order to develop a BIM/GIS digital model.

Deconstruction, section 7.3.1: The often synonymously used terms disassembly, decommissioning, reverse engineering and deconstruction[1] have the aim to "eliminate demolition as an end-of-life option" (Akinade et al. 2017a, p. 261) through the recovery of reusable materials (Gorgolewski 2006). In the best case, no materials are landfilled – but in reality, this is hardly possible (Akinade et al. 2015, p. 168).

Destruction/creation of competencies, section 1.1.1: The central processes of creation of disruptive innovation in the firm, described by Tushman and Anderson (1986), who distinguish between two types of innovation, are those built on the destruction of competencies and those built on the growth of competencies. "Innovation based on the increase of skills is built on and reinforces existing skills, know-how. Innovation based on the destruction of skills renders existing skills, know-how and abilities obsolete and outdated" (Gatignon et al. 2002, p. 7, author's translation).

Design-for-deconstruction and design-for-disassembly (DfD) approaches, section 7.3.4.2: The approaches "aim at modifying building design with respect to different criteria that could be helpful with respect to deconstruction, reuse, recycling and waste at the end of a building's lifecycle" (Akinade et al. 2015 2017a, 2017b, 2018; Denis et al. 2018; Akanbi et al. 2019; Charef et al. 2019).

Digital mock-up business process, section 2.2.1: "An extended digital representation of the product used as a platform for product/process development, communication and validation during all phases of the product's life" (author's translation), a definition proposed in the 1990s by

1 In the following, the term "deconstruction" is used.

the consortium of the European AIT – DMU BP project (Advanced Information Technology in Design and Manufacturing). The main mechanisms used to provide each expert with content adapted to their needs are views, configurations and versions.

Digital twin or virtual asset, section 4.2: It is an information system that transforms unstructured data into structured information around which services are developed. The first service is sharing, in all four states of information, as defined in ISO 19650-1 Part 1 (2018). The digital twin emerges for the project owner as one of the deliverables needed to meet the complete lifecycle management of a work. The asset is to be understood in the patrimonial sense: "Item, thing or entity that has a potential or actual value to an organization" (ISO 19650-1, 2018). It is based on an organized progression between data and information, a description and organization of data, within a context, to structure them in the form of information; it is also based on an ontology, or semantic modeling, that is, an explicit and formal specification of knowledge and the interpretation that can be made of it in a given context, and on a distinction and a distribution between conceptual model and semantic model (ontology). Finally, the digital twin is based on interoperability and therefore on standardization.

End of life, section 7.3.1: "End-of-life or 'grave' summarizes the last stage in a product's life cycle. In this life cycle stage, the function of the product at its location/site is terminated (e.g. a building is removed). If a building component/product or a whole part of a building is reused, then it re-enters the circular economy in a new use phase".

Facility management (FM), section 6.3.3: It is defined as "the effective management of place and space, integrating an organization's support infrastructure to deliver services to staff and customers at best value whilst enhancing overall organizational performance" (Royal Institute of Chartered Surveyors 2020, p. 7). "FM consists of several multidisciplinary activities and is essential to achieve success in any project and organization" (Noor and Pitt 2009).

Formal language, sections 5.1 and 5.3: It is a "language for modeling, calculation, and predication in the specification, design, analysis, construction, and assurance of hardware and software systems whose syntax and semantics are defined on the basis of well-established mathematical concepts" according to ISO/IEC 29128:2011. A formal language consists of an alphabet (set of elements), rules for determining whether an element belongs to the alphabet of the language (grammar), along with a compound set of elements that respects these rules (syntax). A formal language has a meaning or semantics. The more explicitly and logically its meaning is defined, the more formal the language is. The expressiveness of a formal language is the number of operators or constructors allowed to combine concepts to form new ones. Decidability is the ability to apply deduction algorithms to models described with the considered logic language. The more expressive a logical language is, the less decidable it is.

Geographic information systems, section 8.1.1: They are "information systems" (IS) with a "geospatial" (G) component. They include databases (with geo-referenced data or location information), hardware infrastructures (servers, computers, operating system, business software, applications, etc.), a structural organization (management, department, mission, etc.), a human organization (administrator, data scientist, engineers, cartographer, etc.), and an overall governance (data management) (Goodchild et al. 2007). This requires the implementation of specific expertise for the production, management, use, updating, analysis and mapping of data, as well as their availability to various users, via web portals, clouds, PDFs, etc.

Geography markup language (GML), section 8.3.2: It is an XML-derived language for encoding, manipulating and exchanging geographic data. GML consists of a set of XML schemas, which define an open format for the exchange of geographic data. GML is used to describe geographic objects, projection systems, geometry (1D, 2D or 3D, vector or raster), topology, time, units of measurement and attributes of geographic objects. On the basis of GML, it is possible to build specific exchange data models for specialized domains ("community standards"), such as CityGML for cities, WaterML for hydrology, IndoorML for indoor navigation, GeoSciML for geology, etc.

Ground radar penetration (GRP), section 7.2.1: It is an instrument used in geophysics to learn about the structure of the ground surface layer using high-frequency electromagnetic waves (Hossain and Yeoh 2018).

Hyperspectral analysis, section 7.2.1: "Hyperspectral imaging techniques can be applied in order to provide both spectral and spatial information of scenes as a set of high-resolution images. Integrating of a 3D point cloud into hyperspectral images would enable accurate identification and classification of surface materials and would also convert the 3D representation to BIM" (Amano et al. 2018).

IFC, sections 3.1 and 3.4: It is an open interoperable standard format, written in a non-encrypted language (therefore readable by a text editor), allowing it to be interpreted by any software, freeing it from the limits imposed by editors and their proprietary formats, the use of which is subject to paying licenses. The reading and understanding of an IFC file is facilitated by a relatively explicit syntax (little compression, little substitution of text by alphanumeric codes, etc.). In the context of ISO standardization, IFC is based on the EXPRESS definition and its data definition language EXPRESS, whose implementation uses the STEP physical file format. The IFC Implementation Guide provides examples of files written using the STEP physical file syntax.

Information, section 4.2: It is a "reinterpretable representation of data in a formalized manner suitable for communication, interpretation or processing. Note 1 to entry: Information can be processed by human or automatic means" (ISO 19650, 2018).

Information container for data drop (ICDD), section 5.3.3: It is the ISO 21597-1 (2020) standard that specifies an ontology for a container for grouping documents and parts of documents (ISO 21597-1, 2020). The first part of this standard defines the container format for storing documents, using RDF, RDFS and OWL standards. The second part provides "means for linking otherwise disconnected data within those documents" (ISO 21597-2, 2020).

Infrastructure, section 3.5.1: An infrastructure is a "set of facilities built on the ground or underground that allow human activities to be carried out across space" (*Dictionnaire de l'urbanisme et de l'aménagement* 2015, author's translation). It is composed of structures such as earthworks, bridges, tunnels, supporting a superstructure (roadway, railroad, etc.). It is often linear and associated with a right-of-way and dependencies (connections to existing structures). An infrastructure strongly impacts its environment.

INSPIRE (INfrastructure for Spatial InfoRmation in the European community), section 5.2.2: The European directive 2007/2/EC takes up the MDA principles to specify interoperability and accessibility of geographic information to support community environmental policies. In the context of INSPIRE, "interoperability" represents compatibility between two systems by allowing them to exchange information so that other systems can understand them (Ansorge et al. 2016). System functionality is specified in the form of a platform independent model (PIM), which is a common data specification for all INSPIRE-compatible systems. The language used to define this PIM is UML. For operational interoperability, translations of the PIM into platform-specific models (PSMs) are performed via automatic procedures, based on languages such as Java, XML Schema or even Python (Ansorge et al. 2016). For greater detail on the conceptual model, see INSPIRE Technical Guidance available online (https://inspire.ec.europa.eu/inspire-technical-guidance/57753).

Integrated Digital Built Environment (IDBE), section 8.3.4: It is a joint working group founded by OGC and buildingSMART International (bSI) in 2014, with the primary objective of developing a common approach for the fundamental concept of alignment between InfraGML and IFC, through a common conceptual model. This work on alignment has notably extended to the representation of geotechnical data and, more broadly, geological observations related to structures.

Integrated engineering, section 2.2.2: It is "an approach that allows for the integrated and simultaneous design of products and related processes, including production and support, intended to allow designers to take into account, from the outset, all phases of the product lifecycle, from its conception to its withdrawal, including quality, costs, deadlines and user requirements" (NF X50-415, 1994, author's translation). It is widely used in the industry.

Intellectual property (IP), section 3.3.2: It is not part of the technical provisions of the contract and must therefore be specified in a contractual document detailing the administrative clauses of the project. Only the data concerning the finished work are affected by these provisions. For example, know-how and construction methods remain the responsibility of the engineering or construction company and therefore remain their property.

Inter-company cooperation, section 1.2.5: The adoption of the BIM approach implies the cooperation of inter-company teams within the project, since the realization of constructions is accomplished through the allocation of contracts; this is the greatest significance of BIM. This means data sharing between teams that may be working on different software chains. In the current exploration phase, we can see that firms are cooperating on a project-by-project basis in order to explore the uses of the BIM approach.

Interoperability, sections 3.1, 3.2 and 4.2: ISO 17261 defines "interoperability" as "the capability of two or more functional units to process data (3.5) cooperatively" (2012). Implementing interoperability means linking two heterogeneous computer systems so that they can collaborate, which implies reciprocal access to their resources. Interoperability necessitates the ability to exchange information between different software in order to avoid re-entering data, which is a source of errors and digital disconnection. It therefore requires exchange formats to describe the information. A standardization of these formats meets the needs of design, modeling and simulation, by establishing an open object-oriented standard, capable of facilitating the exchange of data between software specific to the construction sector. In order to ensure their sustainability, International Alliance for Interoperability (IAI), comprising several software vendors, was created in 1996 to develop the IFC standard. In 2008, in an effort to communicate the goals of the organization, IAI became buildingSMART International (bSI).

IoT (Internet of Things), sections 7.2.1 and 8.4.1: It is a "global infrastructure for the information society enabling advanced services by interconnecting (physical and virtual) things based on existing or evolving interoperable information and communication technologies" (ISO/IEC TR 29181-9:2017, 3.8, included in ISO/IEC 21558-3:2022) Devices that are connected to each other via the Internet can communicate and are responsible for sending and receiving data. IoT and the acquisition of data from multiple sensors related to a building or a territory (energy consumption, air quality, weather, etc.) concerns both GIS and BIM. IoT devices provide real-time data that present a powerful paradigm for Smart City applications to integrate with BIM and GIS. IoT enables technologies such as sensing, communication, identification and recognition technologies, with the ability to share information across platforms through a unified framework. The integration of IoT technology with infrastructure information is complementary and enables innovative applications.

ISO 19650, section 1.5: "Organization and digitization of information about buildings and civil engineering works, including building information modelling (BIM) – Information management using building information modelling" (ISO 19650, 2018, author's translation). It indicates stages that correspond to the technology stages, with the 2016 Level 2 maturity now included in the stages of ISO 19650. It is divided in two parts: Part 1 on concepts and principles, and Part 2 on data and exchanges for the design and construction stages. The ISO 19650 standard refocuses the BIM approach toward a broader conceptual framework of information management to build both the physical work and the digital twin. There are theorical contributions going in this direction too (Succar and Poirier 2020).

Knowledge-based models, section 5.5: Knowledge-based models allow for the modeling of complex systems and rely on the use of formal logic languages to explicitly specify knowledge. Knowledge-based models are defined without any link to a specific computer implementation. A computer can interpret these models and implicit knowledge can be inferred, using reasoning algorithms.

Laser scanning methods, section 7.2.1: These methods directly capture 3D point clouds and differ triangulation, time-of-flight and phase-comparison. Furthermore, they can be differentiated into stationary and mobile systems with different technology readiness levels, commercialization statuses and development perspectives. An overview on the type (table, hand scans, mobile version, terrestrial, car-mounted, drone-mounted), the precision (mm) and ranges (m) is given by Boardman and Bryan (2018).

Last planner system (LPS), section 6.3.2: It "is a collaborative project planning process that involves trade foremen or design team leaders (the last planners) in the planning and control process in greater detail as the time for the work to be done gets closer from master plans to lookahead and weekly plans" (Ballard 2000). "It supports construction production planning and control by providing systematic routines to increase workflow reliability and process stability. In the UK, it is known as Collaborative Planning and, in the USA, it is sometimes called Pull Planning" (Daniel et al. 2017).

Lean construction, section 6.2.2: It can be understood in terms of tools, principles and theoretical foundations. Tools provide the best known and practice-facing aspect of Lean, aiming to reduce waste in processes and generate better value to customers.

Level of Information Need, section 3.2.1: The concept of Level of Information Need evoked in ISO 19650 and detailed in EN 17412 (2021) defines the level of information needed (geometrical detail, associated documentation, actors involved, deadlines, etc.). This standard can be used in the context of contractual or more informational exchanges. It allows for the requesting and requested actors to describe the information, expected and to be produced, respectively, in the same way. Therefore, the need and the response to the need are coherent. It is a method of describing information that facilitates communication and monitoring of the information produced. Contrary to existing documents dealing with "levels of detail, development" or others, this standard does not standardize the content to be exchanged.

Linked data, section 5.3.1: "Linked data" is a subset of Semantic Web principles, aiming at improving data sharing and reusing at a Web platform. It relies upon the principles of the current Web architecture and extends them to describe knowledge.

MADM (multi-attribute design-making), section 7.3.4.1: It involves making decisions (such as evaluation, prioritization, selection) based on available alternatives characterized by multiple, usually conflicting, attributes (Kühlen 2017).

Mechatronics, section 2.5: It is described in the NF E01-010 standard as "an approach aiming at the synergetic integration of mechanics, electronics, automation and computer science in the design and manufacture of a product in order to increase and/or optimize its functionality. The objective of mechatronics is to obtain an added value superior to the simple sum of the added values of the functions taken separately" (2008, author's translation).

Model-based requirements engineering, section 5.2.1: The MBRE approach concerns the engineering and management of requirements based on models and is independent of any computer tool (Holt et al. 2012). It relies on the use of graphical elements to exchange precise and concise requirements. Requirements are considered one at a time, relative to their definitions, and then given "meaning" by placing them in the "appropriate context". A requirement placed in a context (the actors or hierarchical levels) is called a "use case".

Model-based systems engineering (MBSE), section 5.2.1: The MBSE approach is developed within the framework of INCOSE (International Council on Systems Engineering). This approach supports the specification of systems, their requirements, design, analysis, verification and validation. It applies to all phases of the lifecycle of the system under consideration (Long et al. 2011).

Model-driven architectures (MDA), section 5.2.1: These architectures define rules for structuring the representation of systems in the form of models. Specified in the framework of the Object Modeling Group (OMG), and taking up the principles of model-driven engineering (MDE), MDA approaches (Miller et al. 2003) consider three levels of abstraction (or points of view) through which a system can be described. In the MDA approach, a viewpoint is "an abstraction technique for focusing on a particular set of concerns within a system while suppressing all irrelevant detail. A viewpoint can be represented via one or more models" (Tuyen 2006). The MDA approach defines three main viewpoints corresponding to three types of models: CIM, PIM and PSM.

Multi-mode resource constraint project scheduling problem (MRCPSP), section 7.3.4.1: It is presented as a generalization of RCPSP with scheduling. Activity preparation times and due dates are taken into account in order to study the impact on project lifetimes (Volk 2017; Volk et al. 2018).

Object, section 5.2: "A discrete entity with a well-defined boundary and identity that encapsulates state and behavior; an instance of a class" (Rumbaugh et al. 1999, p. 378).

Object-oriented analysis, section 5.2: The term was first suggested by Coad et al. (1990) to integrate services and messages (concepts from object-oriented programming) in {Entity/Relationship} models. The idea is to improve inheritance management. In this view, an "object" is "an abstraction of something in a problem domain, reflecting the capabilities of a system to keep information about it, interact with it, or both" (Coad et al. 1990). According to this definition, an object-oriented analysis targets the problem space (e.g. the business, the organization, the business domain under consideration) and not the solution space (e.g. computers, languages and computer programs).

Object models, section 5.5: Approaches based on object models allow for the modeling of complex systems and have a strong link with the computer implementation of the system under consideration. They are said to be semi-formal, since they are a (graphical) representation of a (computer) implementation of a complex system. The knowledge (i.e. the interpretation to be associated with the modeled information) is not modeled explicitly, it is implied. The correct interpretation of such an object model requires a human expert.

Observations and measurements, section 8.3.2: O&M is a generic model proposed by the OGC to describe observations and measurements based on a set of specifications to standardize access to sensor data, whether it is a thermometer, a probe measuring the depth of water in a borehole or a satellite observation. This model is also applicable to observations made by a "human" sensor (biodiversity, geology, etc.). O&M is adopted in many environmental disciplines.

OGC standards, section 8.3.1: These standards are key elements of geospatial communication interfaces, encodings, and best practices for sensor networks, location-based services, terrestrial imagery networks, climate models, climate management programs, disasters and national spatial data infrastructures worldwide. Since the early 2000s, the emergence of spatial data infrastructures (SDIs) deployed at regional, national, continental, or international scales has also benefited from OGC standards and encouraged their deployment (Nebert 2001). In Europe, the INSPIRE directive, which defines the rules for building a European Spatial Data Infrastructure, is based primarily on OGC standards. To ensure the coherence of the Internet and Web ecosystems, OGC has partnered with many other standards development organizations and industry associations. They work closely with OGC on a wide range of topics such as integration of indoor/outdoor locations, sensor fusion, urban modeling, location-based marketing, aviation, meteorology, Internet of Things, points of interest and the semantic web. Industrial or technical organizations thus seek to guarantee that their data will be accessible to other communities through OGC standards and that they can, in return, benefit from the data of other communities.

Ontology, section 5.3.1: Formal, explicit specification of a shared conceptualization. Note 1: An ontology typically includes definitions of concepts and specified relationships between them, set out in a formal way so that a machine can use them for reasoning. Note 2: see also ISO/TR 13054:2012, 2.6; ISO/TS 13399-4:2014, 3.20; ISO 19101-1-2014, 4.1.26; ISO 18435-3:2015, 3.1; ISO/IEC 19763-3:2020, 3.1.1.1.

Open GIS Consortium, now the Open Geospatial Consortium Inc. (OGC), section 8.3.1: It was launched in 1994. A not-for-profit, international organization, the consortium brings together stakeholders involved in the development and promotion of open standards that ensure interoperability in the field of geomatics and geospatial information through cooperation among developers, suppliers and users. The OGC has grown from 20 members in 1994 to 520 in 2020, bringing together governmental, academic, and private sector organizations and sectors. Although founded in the United States, the OGC now has more members in Europe than anywhere else (Reichardt and Robida 2019).

OpenBIM, 1.4, sections 3.2 and 3.2.1: Based on the application of standards, it is an implementation of BIM concepts, principles and methods that are operationalized outside of any proprietary format, relying solely on business principles and open standards shared in a free manner. OpenBIM is a robust ecosystem for the development of a BIM approach, throughout the lifecycle of the works and applicable in any type of project and company, whether for the design, construction or operation of the works. OpenBIM is defined by a model-based approach (e.g. IFC) that organizes data and describes the digital asset and a process-based approach, using, for example, ISO 19650 (2018) that follows standardized contract processes. OpenBIM became possible in 2020 when the standards production work of the industry players reached the necessary maturity level. IFC 4.3, announced for 2021, and ISO 19650 Part 1 and Part 2 (2018) began appearing in tenders as early as 2020. OpenBIM as a collaborative approach applicable to the entire lifecycle is based on open standards and work processes. Combined with other standards such as PLCS (Project Lifecycle Support), OpenBIM can be the basis of the work's modeling. Since no single software covers all modeling and simulation needs, it is unavoidable that one needs to use specialized business software.

OWL languages, section 5.3.1: OWL languages (Bechhofer et al. 2004) use languages from the Description Logics (DL) language family to axiomatize knowledge of a given business domain. Depending on the implemented DL constructors, different OWL profiles are specified by the W3C. In order to identify OWL terms, the prefix "owl" is used; it corresponds to the domain name "http://www.w3.org/2002/07/owl#". The OWL family of languages defines two types of properties: the so-called object properties (owl:ObjectProperty) and the so-called data type properties (owl:DatatypeProperty). The latter are used in triples that have a value or a datatype as object, while object properties are used in triples that have classes or instances of classes as object.

Parametric design, section 6.3.1: In parametric design, the user defines associations between design elements and geometries – "this window depends upon this wall and will move with it" (Vermeulen and Ayoubi 2019, author's translation).

Photo- and videogrammetry, section 7.2.1: Photo- and videogrammetry construct 3D point clouds and 3D models from multiple images. Methods include close-range (<300 m) and aerial (>300 m) photogrammetry. While close-range photogrammetry is more suitable for capturing buildings and infrastructures, aerial methods are applied for survey, mapping and cartography. Both types become indistinct by using drone-mounted cameras, for example, those based on DJI drones (DJI 2019), with lower flight altitude. Photogrammetry uses the concept of stereoscopy to detect and extract spatial distances and relations from multiple, partly overlapping images. Factors such as image resolution, lighting conditions and distance from the object influence the 3D model quality. Free and commercial solutions for respective image processing include, for example, Autodesk® ReCap™ (Autodesk 2019), DroneDeploy or Pix4D (All3DP 2019).

Plan-Do-Check-Act (PDCA), section 6.4: The PDCA cycle offers itself for evaluating designed solutions and for continuous improvement.

Platform independent model (PIM), section 5.2.1: PIM describes the operational capabilities of a system (defined as abstractions of the platform), regardless of platform (or set of platforms) specific implementation details.

Platform specific model (PSM), section 5.2.1: PSM represents a translation of a PIM with respect to a specific platform. These PSMs are defined through so-called implementation languages (e.g. Java, Python or XML Schema). Automated tools (e.g. model transformation tools) are used to ensure the translation of the PIM into different PSM models.

Product lifecycle management (PLM), sections 2.1 and 2.3.1: It is a business strategy that aims to create, manage and share all the information about the definition, manufacture, maintenance and recycling of an industrial product, throughout its lifecycle, from the preliminary studies to the end of its life (Amann 2002).

Product process organization (PPO), section 2.4.1: It proposes a product model (function, behavior, structure) and links it to expert tools (e.g. CAD), a process model for tracing and capitalizing on knowledge evolutions, and an organization model facilitating multi-objective decision-making. Developed in the RTNL IPPOP project (Integration of Product Process Organisation for engineering Performance improvement) (Robin et al. 2007; Noël and Roucoules 2008).

Project management, section 1.2.3: According to Morris (2013), project management is "a set of concepts, tools, and techniques on how to execute projects on time, within budget, and according to required client specifications within the context of an explicit business strategy" (p. 63, author's translation). With BIM, project management is profoundly impacted through the processes of collective elaboration and exchange. Project management is still of the control type but renewed by the recomposition imposed by the central use of digital technology.

Project management plan (PMP), section 4.2.2: It is the system "for making it", including the concepts of system engineering and requirements engineering (product, pro-process, modeling or information) that allow users to structure the BIM process and the other processes of the project.

Project portfolio and corporate strategy, section 1.2.4: On the one hand, it is a matter of capitalizing from one project to another (Loufrani-Fedida and Missonier 2015), but moreover, of determining strategies that cut across projects, of conceiving the set of projects as a portfolio that explores, in a differentiated way, the global technological mutation to be made. The strategy cannot be only capitalization and bottom up; it must also be transversal and top down, in order to allow for centralized strategic thinking and decision-making. It is therefore not a logic of accumulation that must be implemented, but rather a logic of strategic orientation.

Pull technique, section 6.3.3: In the context of FM it means that only such information should be produced upstream that the customer downstream needs to operate and maintain the facility (Succar 2009).

Quality management maturity grid (QMMG), section 1.5: A maturity model (MM) proposed by Crosby (1979), the QMMG, described as a descriptive maturity grid or matrix (Fraser et al. 2002), contains textual descriptions for each activity at each maturity level.

Radio frequency identification (RFID), section 7.2.1: It has emerged as an automatic data collection and information storage technology and has been used in different applications in the AEC/FM (architecture, engineering, construction, and facilities management) industry. RFID tags are attached to building assets throughout their lifecycle and used to store lifecycle and context-aware information taken from a BIM.

RDF model, section 5.3.1: According to the W3C, "RDF is a standard model for data interchange on the Web. RDF has features that facilitate data merging even if the underlying schemas differ, and it specifically supports the evolution of schemas over time without requiring all the data consumers to be changed. RDF extends the linking structure of the Web to use URIs to name the relationship between things as well as the two ends of the link (this is usually referred to as a 'triple'). Using this simple model, it allows structured and semi-structured data to be mixed, exposed, and shared across different applications. This linking structure forms a directed, labeled graph, where the edges represent the named link between two resources, represented by the graph nodes. This graph view is the easiest possible mental model for RDF and is often used in easy-to-understand visual explanations" (source: https://www.w3.org/RDF/).

RDF schema, section 5.3.1: According to the W3C, "RDFS is a general-purpose language for representing simple RDF vocabularies on the Web. Other vocabulary definition technologies, such as OWL or SKOS, build on RDFS and provide language for defining structured, Web-based ontologies which enable richer integration and interoperability of data among descriptive communities" (source: https://www.w3.org/2001/sw/wiki/RDFS). Brickley et al. (2014) introduce a vocabulary on top of RDF, with different terms to further characterize statements formed with RDF. To identify RDF Schema terms, the prefix used is rdfs corresponding to "http://www.w3.org/2000/01/rdf-schema#". With RDF Schema, it is possible to define classes (rdfs:Class), subclasses (rdfs:subClassOf), sub properties (rdfs:subPropertyOf) as well as starting (rdfs:domain) and ending (rdfs:range) sets for properties.

Recycling, section 7.3.1: It is the reuse of building materials, components and products after their structure has been destroyed and post-processed into a new component or product made from only a secondary or partially secondary raw material (building a secondary mine or material bank).

Scan-to-BIM, section 7.2.1: "Scan-to-BIM is the process of 3D conversion of the structure to be rehabilitated into a digital representation (BIM)" (Czerniawskia and Leite 2018, author's translation). It is a process of surveying the existing structure with a 3D laser scanner. This process allows for the digital acquisition of millions of reference points of an existing structure, in the form of a point cloud. The digitized data are then imported into a 3D modeling environment to reconstruct a digital as-built model of the structure.

Semantic modeling, section 5.3: In semantic modeling, an "entity" is not concerned with operations, methods or behavior (unlike an "object" in the "object-oriented" world). These elements belong to the domain of "process modeling". Knowledge-based models use the concept of "class" differently from object-oriented models. In semantic modeling, a "class" groups together instances that have the same interpretation and share a set of properties.

Semantic Web, section 5.3.1: As with the traditional Web, the Semantic Web relies on the use of URIs as a mechanism for unique and global resource identification, at a Web scale. The http protocol is, as for the traditional Web, the universal resource access mechanism. The difference lies in the description of the resources. While the traditional Web uses HTML for Web pages, the Semantic Web relies on RDF (Hayes et al. 2014) for resource description (thus overcoming limitations of <a href> links used in HTML). See RDF model.

Sequence disassembly planning for buildings (SDPB), section 7.3.4.2: This method seeks to minimize environmental impact and disassembly costs by using recursive analyses based on defined rules to plan the recovery of target components of multi-instance building subsystems according to physical, environmental and economic constraints.

Shape grammar, section 7.2.2: The "method uses logical rules to infer, predict and model interior spaces from missing or incomplete data in 3D parametric IFC models" (Tran et al. 2018). Furthermore, topological relations (e.g. containment, adjacency and connectivity) are derived. The resulting BIM has high geometric accuracy and rich semantic content.

Static or as-built model, section 1.5: The static as-built model is opposed to the as-built model.

STEP (Standard for the Exchange for Product model data), sections 2.4.1 and 5.2.2: STEP (ISO 10303) is the most widely used standard for the data exchange on the entire product lifecycle (Rachuri et al. 2008). Developed under the responsibility of ISO, within the TC184/SC4 subcommittee "Industrial data", this standard specifies, among other things, several Application Protocols (AP) that define several specialized data models for one or more domains.

STEP AP 225 protocol, section 5.2.2: It describes the requirements associated with building elements, and the STEP AP 242 protocol specifies the requirements associated with managed model-based 3D engineering. Under these protocols, needs and requirements are specified in natural language, according to the Information Delivery Manual or IDM standard (ISO 29481-1, 2016). According to the IDM standard, the formal syntax to be used for construction data exchange is that specified by the Industrial Foundation Classes or IFC (ISO 16739-1, 2018): it is either the EXPRESS format, as defined in (ISO 10303-21, 2016), or the XML format.

System, section 4.2.1.1: A "system" is defined as "a set of interdependent or temporally interacting parts" (Levin 2006), where the parts are, in general, themselves systems comprising other parts.

System "for making" and system "to be made", section 4.2.2: ISO 19650 (2018) structures the object to be realized (the product or system "to be made") and the processes, ensuring the lifecycle of the data and its qualification (the project or system "for making it"). The use of system engineering and requirements engineering is reinforced by the use of common modeling formalisms (Tolmer 2016). The digital mock-up of a work carrying a certain value, complementary to that of the physical work itself, implies working on the organization of the system "for making it" as much as on the management of the complexity of the system "to be made", often better controlled. The complexity of the works is generally technically better mastered than the complexity of the exchanges and sharing of information, even for the most complex works. The differentiation of the system "to be made" and the system "for making it" facilitates the management of these two project complexities. However, the use of these concepts and methodological tools only benefits the system "to be made" if they are mobilized by and for the system "for making it". The verification of requirements is carried out directly on the system. The verification of the requirements takes place directly on the system "to be made" during the acceptance of the works.

Systems engineering (SE), section 2.6: It is "a general methodological approach that encompasses the set of activities appropriate for designing, evolving, and verifying a system that provides a cost-effective and efficient solution to a customer's needs while satisfying all stakeholders" (AFIS 2012, author's translation). It is a recognized approach to support multidisciplinary product development and is generally associated with the product development process known as the V-cycle (Dieterle 2005; Kleiner and Kramer 2013).

Target value design (TVD), section 6.3.1: It is a collaborative design development process focused on delivering project value for or below a targeted project cost.

Technological disruption, section 1.1: A disruptive innovation or disruptive technology is a technological innovation that involves a product or service that ends up replacing a previously dominant technology in the market (Bower and Christensen 1995; Christensen et al. 2015). Disruptive technologies, unlike continuity technologies, impact markets and create many discontinuities, both in customer segments, in the organization and industry and in the firm.

Transformation, flow and value (TFV) Theory, section 6.3.3: "The Transformation, Flow and Value (TFV) theory of production (Koskela 2000) has been used to explain the difficulties in implementing BIM in FM (Shou et al. 2014, Munir et al. 2019). From the TFV theory perspective, even though the focus of FM has shifted from cost minimization in real estate operations to supporting end-customer requirements using BIM, the existing literature on BIM-enabled FM does not explain the customer value generation process or information flows between stakeholders" (Munir et al. 2019).

V-cycle, section 2.6: The V-cycle is the representation of a process, very generalized, which starts with the identification of user needs and ends with the validation of the user. It is broken down into two main phases: the so-called top-down phase of decomposition and definition of the product and the so-called bottom-up phase of integration and recomposition.

Virtual reality (VR)/mixed reality (MR), section 6.3.1: Immersive technologies such as virtual reality (VR)/mixed reality (MR) are now more frequently used with BIM in client and stakeholder engagement for the communication of design intent and requirement capture as well as coordination and verification through advanced visualization (Zaker and Coloma 2018, Wang et al. 2018, Ergün et al. 2019).

Web map service (WMS), section 8.3.2: This standard communication protocol makes it possible to obtain georeferenced data maps from different data servers distributed on the Web (De La Beaujardière 2002). This allows users to set up a network of map servers from which clients can build interactive maps that can overlay data from different origins and of different natures. Changes of the projection systems are taken into account by the service.

List of Authors

Marie BAGIEU
ESITC Caen
France

Pierre BENNING
Bouygues – Travaux Publics
and
Projet MINnD
Paris
France

Matthieu BRICOGNE
Roberval
Université technologique de Compiègne
and
Groupement francophone des utilisateurs de Catia
Paris
France

Christophe CASTAING
buildingSMART International
and
Groupe Egis
and
Projet MINnD
Paris
France

Claude DUMOULIN
Bouygues – Travaux Publics
and
Projet MINnD
Paris
France

Alexandre DURUPT
Roberval
Université technologique de Compiègne
France

Benoît EYNARD
Roberval
Université technologique de Compiègne
and
Inetum
Meudon
France

Hervé HALBOUT
HALBOUT Consultants
Caen
and
Projet MINnD
Paris
France

Mojgan A. JADIDI
Lassonde School of Engineering
York University
Toronto
Canada

Arto KIVINIEMI
School of Architecture
University of Liverpool
United Kingdom

Lauri KOSKELA
School of Arts and Humanities
University of Huddersfield
United Kingdom

Julien LE DUIGOU
Roberval
Université technologique de Compiègne
France

François ROBIDA
BRGM : Service géologique national français
Orléans
and
Projet MINnD
Paris
France

Ana ROXIN
Laboratoire d'Informatique de Bourgogne
Université Bourgogne-Franche-Comté
Dijon
France

Saeed TALEBI
School of Engineering and Built Environment
Birmingham City University
United Kingdom

Régine TEULIER
CNRS, CRG
Institut polytechnique
and
Projet MINnD
Paris
France

Algan TEZEL
College of Engineering & Physical Sciences
Aston University
Birmingham
United Kingdom

Charles-Édouard TOLMER
Eurovia Infra
Vinci Construction
Nanterre
and
Projet MINnD
Paris
France

Patricia TZORTZOPOULOS
School of Arts and Humanities
University of Huddersfield
United Kingdom

Rebekka VOLK
Institute for Industrial Production
Karlsruhe Institute of Technology
Germany

Index

D

E, F, G

I, L

M, O

P, Q

R, S, T

U, V, W, X

Printed and bound by CPI Group (UK) Ltd, Croydon, CR0 4YY

13/09/2026

14960998-0002